E. Seibold   W. H. Berger

The Sea Floor

An Introduction to Marine Geology

With 206 Figures

Springer-Verlag
Berlin Heidelberg New York 1982

Professor Dr. Eugen Seibold
Deutsche Forschungsgemeinschaft
Kennedy-Allee 40
D-5400 Bonn-Bad Godesberg
FRG

Dr. Wolfgang H. Berger
Scripps Institution of Oceanography
Geological Research Division
La Jolla, California 92093
USA

ISBN 3-540-11256-1 Springer-Verlag Berlin Heidelberg New York
ISBN 0-387-11256-1 Springer-Verlag New York Heidelberg Berlin

Library of Congress Cataloging in Publication Data.
Seibold, Eugen. The sea floor.
Bibliography: p. Includes index. 1. Submarine geology. I. Berger, Wolfgang H.
II. Title.
QE39.S38   551.46'08   81-18434
ISBN 0-387-11256-1 (U.S.)   AACR2

Typesetting, printing and binding: Kösel GmbH & Co., Kempten
2132/3321/543210

Contents

Introduction

Man's understanding of how this planet is put together and how it evolved has radically changed during the last twenty years. This great revolution in geology — now usually subsumed under the concept of Plate Tectonics and its ramifications — was an outgrowth of the study of the ocean floor. In its impact on the earth sciences, it is comparable to the revolution brought about by Charles Darwin (1809–1882) a century ago. The evolution of the biosphere was the focal point at that time. We are now discussing the evolution of the lithosphere, the uppermost 100 km of the Earth. Darwin drew his inspiration from observations made during the voyage of the Beagle, and his work gave strong impetus to the first global oceanographic expedition, the voyage of HMS Challenger (1872–1875). Ever since, oceanographic research has been intimately associated with fundamental advances in the knowledge of Earth. This should come as no surprise. After all, our planet's surface is mostly ocean.

This booklet is the result of our conviction that to study introductory geology and oceanography and environmental science, one needs a summary of the tectonics and morphology of the sea floor, of the geologic processes active in the deep sea and in shelf seas, and of the climatic record in deep sea sediments.

Our aim is to give a brief survey of these topics. We have endeavored to write for all who might be interested in the subject, including those without formal training in the natural sciences. The coming decades will be characterized by an increasing awareness of man's dependency on natural resources, including the ocean as a weather machine, a waste bin, and a source of energy and minerals. An important part of this awareness will be an appreciation for the elementary facts and concepts of marine geology.

In what follows we shall first give a brief overview of the effects of endogenic forces on the morphology of the sea floor. Several excellent summaries for the general reader are available on this topic which is closely linked to the theory of continental drift, and has been a focus of geologic discussions for the last two decades. For the rest, we shall emphasize the exogenic processes, which determine the physical, chemical, and biological environment on the sea floor, and which are especially relevant to the intelligent use of the ocean and to an understanding of its role in the evolution of climate and of life.

The results and ideas we report are the product of the intensive work by many dedicated marine geologists. We introduce some particularly distinguished scientists by portrait (Fig. P.1). There are many more, and we have occasionally mentioned the authors of important contributions. However, we did not find it possible in a book like this to give credit systematically where credit is due. We sincerely apologize to our colleagues for this unscholarly attitude, citing necessity in defense. For those who wish to pursue in greater depth the subjects discussed, we append a list of key references.

The present booklet was developed from an earlier one by E. S. *(Der Meeresboden)*, which serves as a supplementary college text in Germany. E. S. rewrote parts of it. W. H. B. translated and edited the text for this English edition, and added material on deep sea sedimentation, Pleistocene oceanography, and on the results of the Deep Sea Drilling Project. The senior author thanks his collaborators at the Institute for Geology and Paleontology at the University of Kiel, who have shared the excitement of discovery on numerous expeditions of the research vessel METEOR and other ships, for the last two decades. The junior author owes similar debts to his teachers, colleagues, and students at Scripps Institution of Oceanography, in La Jolla, California. Special thanks to our friends and colleagues who helped us assemble the portraits for Fig. P.1.

Kiel and La Jolla, Spring 1981   E. Seibold
                                                                  W. H. Berger

Fig. P.1. Founders of marine geology. Upper row (left to right): John Murray (1841–1914), Johannes Walther (1860–1937), Alfred Wegener (1880–1930). Middle: Jacques Bourcart (1891–1965), Philip H. Kuenen (1902–1976), Harry H. Hess (1906–1969). Bottom: Maurice Ewing (1903–1976), Bruce C. Heezen (1924–1977), Francis P. Shepard (born 1897)

3

Pioneers of Marine Geology (see Fig. P.1)

Marine geology is a young offshoot of geology, a branch of science which begins with James Hutton (1726–1797) and his *Theory of the Earth* (Edinburgh, 1795). Among other things, Hutton studied marine rocks on land. Changes of sea level ("encroachment of the ocean", and "the placing of materials accumulated at the bottom of the sea in the atmosphere above the surface of the sea") was a central tenet of his "Theory". Thus, the question of what happens at the bottom of the sea was raised at the beginning of systematic geologic investigation. This question had to be attacked if the marine deposits on land were to be understood. Hutton was not alone in these concerns. A few years before the *Theory of the Earth* appeared, the great chemist Antoine Laurent Lavoisier (1743–1794) distinguished two kinds of marine sedimentary layers, that is, those formed in the open sea at great depth, which he called *pelagic* beds, and those formed along the coast, which he termed *littoral* beds. "Great depth" for Lavoisier was everything beyond wave base. Lavoisier supposed that sediment particles would settle quietly in deep water and reworking would be much less in evidence here than near the shore.

Marine geology starts in earnest with geologists going to sea and looking for the processes which helped produce the marine rocks with which they were familiar on land. Such work, obviously, began in the intertidal and in easily accessible shallow waters, and it was undertaken by a great many investigators. The German geologist Johannes Walther (Fig. P.1) was very familiar with the observations resulting from this work and was himself a pioneer in these types of studies. Firmly rooted in classical geology, he set an outstanding example of applying *uniformitarianism,* that is, Hutton's doctrine that observable processes are sufficient to explain the geologic record (*Lithogenesis of the Present,* Jena, 1894, in German). In his book *Bionomie des Meeres* (Jena, 1893), he gave an account of marine environments and the ecology of their inhabitants, with emphasis on shell-forming organisms and the deposits they produce. Research on marine sedimentation in the following four decades is summarized in P. D. Trask (ed) *Recent Marine Sediments* (AAPG, Tulsa, 1939). This symposium spans the range of sedimentary environments, from beach to deep sea, and many of the articles were written by pioneers in marine geology.

As marine geological studies progressed and moved further out to sea, there was a gradual change of emphasis in the set of problems to be solved. The seafloor itself became the focus of attention, not for the sake of the clues it would yield for the purposes of land geology, but for the clues it contained for its own evolution and its role in the history of Earth. The new emphasis is first evident in the works of the Scotsman, John Murray (Fig. P.1), naturalist on the HMS Challenger Expedition (1873–1876). The expedition, led by the biologist Charles Wyville Thomson (1830–1882), marks the beginning of modern oceanography. It established the general morphology of the deep sea floor and the types of sediments covering it. John Murray's chief opus, *Deep Sea Deposits* (written with A. F. Renard and published in 1891) laid the foundation for the sedimentology of the deep ocean floor (Chap. 3). Murray's studies established a fundamental dichotomy of shallow water and shelf-sediments on the one hand, and of deep sea deposits on the other, and it became a commonplace textbook truism that no true deep sea deposits are found on land anywhere. This doctrine was challenged when Ph. H. Kuenen (Fig. P.1) demonstrated by experiment that clouds of sediment could be transported downslope on the seafloor at great speed and to great depth, due to the fact that muddy water is heavier than the clear water surrounding it. Sediment transported in suspension in this fashion settles out at its site of deposition, heavy and large grains first, fine grains last. The resulting layer is *graded,* and such layers are indeed common in the geologic record (e. g. Alpine *flysch* deposits). Strong circumstantial evidence that Kuenen's concept is important in deep sea sedimentation was first presented by B. C. Heezen and M. Ewing, in 1952 (see Chap. 4, 3.6). Kuenen made many other important contributions to marine geology, attacking a broad range of subjects in his book, *Marine Geology* (New York, 1950), and in numerous publications.

Studies in marine sedimentation eventually led into ocean history when geologists started to take cores. The pioneering expeditions were those of the German vessel Meteor (1925–1927), which first established deep sea sedimentation rates and of the Swedish vessel Albatross (1947–1948), led by Hans Pettersson. The Albatross results established the presence, in all oceans, of cyclic sedimentation due to the climate fluctuations during the last million years, which included several ice ages (Chap. 9). The last (and biggest) effort in this line is the ongoing Deep Sea Drilling Project, using the vessel Glomar Challenger (1968-present), which first made Tertiary and Cretaceous sediment sequences available for systematic study.

The investigation of the morphology of the seafloor proceeded parallel to that of marine sedimentation. Coastal landforms were the most accessible, and considerable information had been accumulated early in this century (D. W. Johnson, *Shore Processes and Shoreline Development,* Wiley, New York, 1919). Much additional work on these topics was done by F. P. Shepard (Fig. P.1) and also by J. Bourcart (Fig. P.1), who were able to test earlier concepts against field data collected in shallower water. These two pioneers of marine geology especially attacked the problems of the origin of continental margins, and of submarine canyons, by studying their morphology and associated sedimentary processes.

F. P. Shepard's wide-flung field areas include the shelf off the U.S. East Coast, shelf and slope off the U.S. West Coast, and the floor of the Gulf of Mexico. His textbook *Submarine Geology* (New York, 1948) and subsequent editions summarize the results of this work and present global statistics on sea floor morphology. Other representative works are *Recent Sediments, Northwest Gulf of Mexiko* (Tulsa, 1960, with F. B. Phleger and Tj. H. van Andel) and *Submarine Canyons and Other Sea Valleys* (Chicago, 1966, with R. F. Dill). J. Bourcart carried out similar geomorphologic and sedimentologic studies off the shores of France, especially in the Mediterranean. His concept of continental margin "flexure", with uplift landward of a *hinge-line* and downwarp seaward of it, proved useful in explaining the migrations of sea level across the shelf, and in studying the nature of sediment accumulation on the continental slope (see Chap. 2).

The marine geomorphologist par excellence was B. C. Heezen (Fig. P.1), whose physiographic diagrams of the seafloor (drawn with his collaborator, Marie Tharp) show great insight into the tectonic and sedimentologic processes of the seafloor. His graphs now appear in virtually all textbooks of geology and geography (see Figs. 1.3 and 2.2). Of B. C. Heezen it has been said (somewhat facetiously, by E. C. Bullard) that he perfected the art of drawing maps of regions where no data are available. Much of his work is summarized in the beautifully illustrated book, *The Face of the Deep* (New York, 1971, with C. D. Hollister).

A satisfying explanation of the overall morphology of ocean basins and of the various types of continental margins could only come from geophysics, which deals with the motions and forces deep within the earth. It is no coincidence that a geophysicist first formulated a global hypothesis for ocean-margin morphology which proved to be viable: the meteorologist Alfred Wegener (Fig. P.1).

A. Wegener became intrigued with the parallelism of the coast lines bordering the Atlantic Ocean (a phenomenon noted already in 1801 by the famous naturalist-explorer, Alexander von Humboldt). It seemed to Wegener that the continents looked like puzzle pieces which belong together. He then learned quite by accident that paleontologists had invoked former land bridges between the shores facing each other across the Atlantic to explain striking similarities between the fossil records of both sides of the ocean. After an extensive literature search he became convinced that the continents were once joined, and broke up after the Paleozoic. Replacing the concept of land bridges with his hypothesis of continental drift, he started the "debate of the century" in geology with his book, *The Origin of Continents and Oceans* (Braunschweig, 1915, in German). His hypothesis is now an integral part of *sea floor spreading* and *plate tectonics* (Chap. 1), the ruling theories explaining the geomorphology and geophysics of the ocean floor and of the crust of the Earth in general (Chap. 1). It was the work of seagoing geophysicists which eventually led to the acceptance of *plate tectonics*. The pioneering efforts of E. C. Bullard (1907–1980) on earth magnetism, heat flow, and seismic surveying, and of M. Ewing (Fig. P.1) and his associates in all aspects of marine geophysics were of central importance in this development (although M. Ewing did not himself advocate sea floor spreading). These scientists also were instrumental in elucidating the structure of continental margins (M. Ewing et. al., 1937, Geophysical Investigations in the Emerged and Submerged Atlantic Coastal Plain, Bull Geol Soc Am, 51, p 909; E. C. Bullard and T. F. Gaskell, 1941, Submarine Seismic Investigations, Proc Royal Soc, Ser A 177, p 476).

The turning point in the scientific revolution which shook the Earth Sciences, and which culminated in *plate tectonics* is generally taken to be the seminal paper of H. H. Hess, *History of Ocean Basins,* published in 1962. Hess (Fig. P.1) started his distinguished career working with the Dutch geophysicist F. A. Vening-Meinesz, on the gravity anomalies of deep sea trenches. These investigations resulted in the hypothesis that trenches may be surface expressions of the downgoing limbs of mantle convection cells. As a naval officer, Hess discovered and mapped

6

a great number of flat-topped seamounts, whose morphology suggested widespread sinking of the sea floor. Stimulated by subsequent discoveries on the Mid-Ocean Ridge (rift morphology, heat flow, and others) he proposed his idea of the sea floor being generated at the center of the Mid Ocean Ridge, moving away and downward as it ages, finally to disappear into trenches. The term "sea floor spreading" was introduced by R. Dietz (in 1961) for this postulated phenomenon. *Sea floor spreading,* and its offspring *plate tectonics* since became the basic framework within which the data of marine geology are interpreted. Chap. 1 gives background and details.

1 Origin and Morphology of Ocean Basins

1.1 The Depth of the Sea

The obvious question about the sea floor is to ask how deep it is and why. The overall depth distribution first became available through the voyage of H. M. S. Challenger (Fig. 1.1). We see that there are two most common depths: a shallow one near sea level (shelf seas), and a deep one between 4 and 5 km (normal deep ocean). The sea floor connecting shelves and deep ocean is of intermediate depths and makes up the continental slopes and rises. There is a portion of sea floor which is twice as deep as normal: such depths only occur in narrow trenches, mainly in a ring around the Pacific Ocean (Appendix A2).

On H. M. S. Challenger, depth soundings were done by laboriously sending a weight to the ocean floor and measuring the length of the wire paid out. When the scattered soundings were connected in drawing depth contours, the ocean floor looked smooth. Only when echo sounding was used routinely did it become

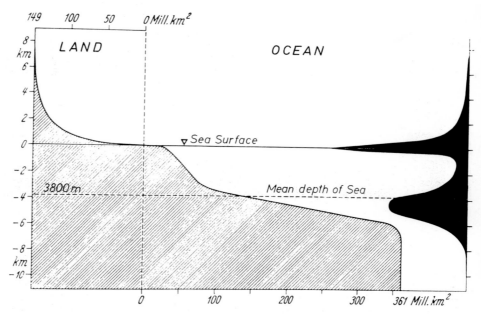

Fig. 1.1. Overall depth distribution of the ocean floor and land elevations (hypsographic curve). *To the right:* frequency distribution of elevations

obvious that large parts of the ocean floor consist of immense mountain ranges whose cragginess rivals that of the Alps and the Sierra Nevada. Perhaps the most impressive of these ranges is the Mid-Atlantic Ridge, first discovered by the famous Meteor Expedition (1925–1927) (Fig. 1.2).

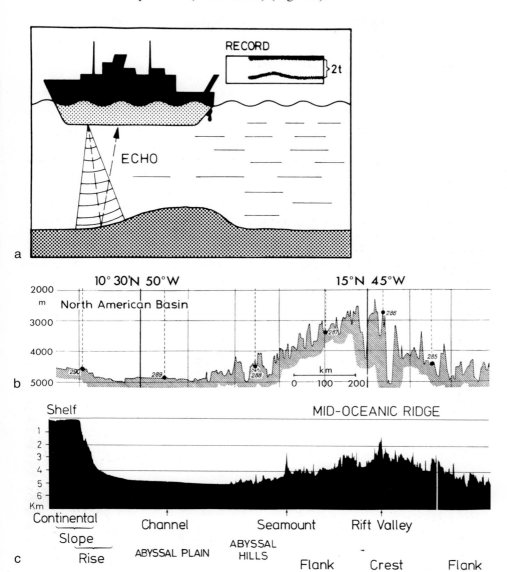

Fig. 1.2a–c. Topography of the Mid-Atlantic Ridge and adjacent sea floor. **a** Principle of continuous echo profiling. The depth is given by $s = v \cdot t$, where v is the sound velocity and 2t is the time the sound takes to travel to and fro. *Record* output of the echo sounding device, on a moving strip of paper. **b** Results of echo sounding as obtained by the German Meteor Expedition (1925–1927). Soundings were taken every 4.5 km. Numbers refer to sampling stations. Note central rift. **c** Modern topographic profile with labels of physiographic features

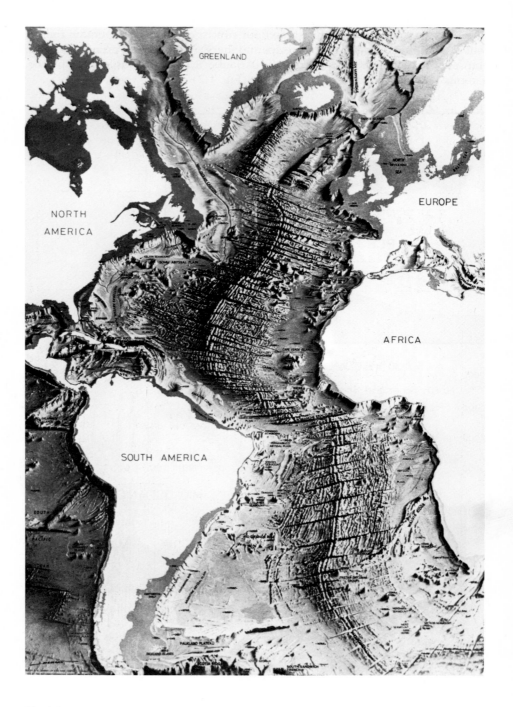

Fig. 1.3. Physiographic diagram of the Atlantic ocean floor, from a painting by H. C. Berann (National Geographic Society), based on bathymetric studies of B. C. Heezen and M. Tharp

10

More recently it has been shown, largely through the work of M. Ewing and that of his associate B. C. Heezen at Lamont Geological Observatory, that the seemingly endless Mid-Atlantic Ridge (Fig. 1.3) is itself only a portion of a world-encircling Mid-Ocean Ridge.

This was, of course, a discovery of immense importance. It identified the one unifying morphological feature of the planet, the central template to which the various scattered puzzle pieces of knowledge about the sea floor had to be fitted. The only other feature of the ocean floor of comparable magnitude is the line of trenches ringing the Pacific (Fig. 1.4). The complementary significance of these

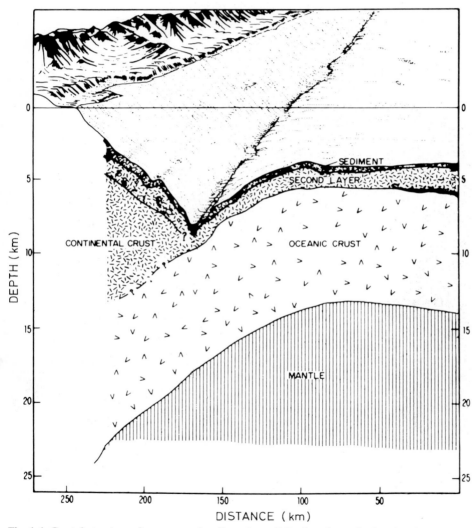

Fig. 1.4. Crustal structure of ocean margin off northern Chile, based on seismic refraction studies. (After R. L. Fisher, R. W. Raitt, 1962, Deep-Sea Res 9: 423)

11

two features — the Ridge System and the Trench System — became obvious in the 1960's, from the study of magnetic properties of the sea floor, from earthquakes, and from heat flow distribution. In the late 1960's the hypothesis that new sea floor forms at the center of the Mid-Ocean Ridge and that it travels toward the trenches where it sinks, gained general acceptance.

The hypothesis, called *sea floor spreading,* explains in an elegant fashion the major features of the depth distribution of the sea floor.

Before we discuss this striking concept of sea-floor spreading in some detail, however, let go back to consider the basic processes which shape Earth's surface, including the floor of the ocean.

1.2 Endogenic Processes

As is true of all of the face of the Earth, the sea floor is shaped by two kinds of processes, those deriving their energy from inside the Earth, called *endogenic,* and those driven by the Sun, called *exogenic.*

The forces inside the Earth produce volcanism and earthquakes; we meet them in the eruptions on Hawaii, in the geysers of Yellowstone Park, in the quakes of California. Working over long periods of time the endogenic forces, called forth by heat sources within the Earth, build mountain ranges such as the Sierra Nevada and the Himalayas, or create gigantic rifts, such as Death Valley and the Rhine Graben. It is reasonable to suppose that the undersea mountains represent uplift, and that the great trenches result from downwarping of the sea floor, by endogenic forces. Such motion, of course, requires flow of material within the Earth. Thus, matter has to rise to make the undersea mountain ranges, and must sink to make the trenches. The mental jump in formulating the hypothesis of sea-floor spreading was to see these necessary motions as part of a convection system (Fig. 1.5).

The exact nature of the forces and motions deep within the Earth is not accessible to direct observation by sampling. Even the deepest drill holes (near 10 km) only scratch the surface of the Earth. They do not penetrate the *crust* which is defined as the uppermost layer of the solid Earth. Continental crust is about 20 to 50 km thick, while oceanic crust is much thinner, roughly 5 km. The *mantle* below the crust is the source of the endogenic forces working on the crust. It is 2850 km thick and makes up about two thirds of the mass of the Earth (Fig. 1.6). The remaining one third is mainly *core* (radius = 3470 km). Only 0.4% of the mass of Earth is in the crust. On a globe 6 ft in diameter, the ocean would be less than 1 mm deep. Even the 100-km-thick *lithosphere,* made up of crust and uppermost mantle and carrying the continents, would be no thicker than the line of a fat marker pen.

Thus, the mountain ranges and great trenches are but small wrinkles on the planet. The motions of sea floor and continents are perturbations on the very surface of a vast globe of hot rock. What kind of rock? And why hot? We do not know for sure. The kinds of rock dredged from deep clefts in the Mid-Ocean Ridge, and those recovered by deep drilling into the Ridge basalt, presumably

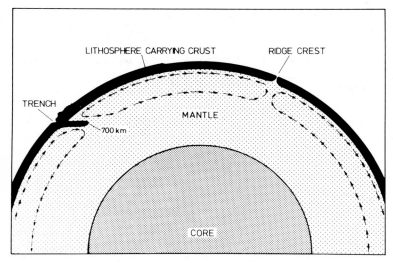

Fig. 1.5. Hypothetical convection currents in the upper mantle, producing sea floor spreading and continental drift. This is one of a number of convection models which have been proposed to explain the origin of mid-ocean ridges and of trenches

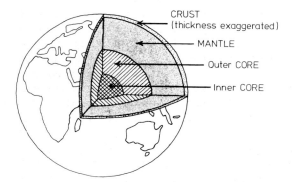

Fig. 1.6. Onion structure of Earth: crust, mantle, and core

most closely resemble the upper mantle material (see Appendix A 7). As mantle material pushes up at the Mid-Ocean Ridge, it changes its character through pressure release, degassing, and especially through reaction with seawater.

The heat most probably comes from the decay of radioactive elements. Another possible source of heat is the gravitational segregation of heavy and light material, which produced the onion structure of Earth in the first place. To receive the "messages from the mantle", that is, information about its structure and the processes in the interior, geophysicists make seismic measurements, and study the characteristics of magnetic and of gravity fields. Mineralogists and petrologists investigate the behavior of rocks under high pressures and temperatures, and geochemists collect indirect evidence on the interior, derived from elemental abundances in the solar system in combination with the density distribution inside the Earth.

13

1.3 Exogenic Processes

Although the large-scale features of the ocean floor are shaped by endogenous forces, the sea floor also reflects the workings of the exogenic forces, that is, the processes of erosion and sedimentation. The classic example is the type of sea floor called abyssal plain — incredibly flat areas hundreds of miles in diameter (Fig. 1.2c). On land, the playas surrounding the Great Salt Lake in Utah can convey a feeling for the nature of these features.

The abyssal plains are vast undersea playas collecting debris from the continents, which is produced by the ever-present agents of weathering: rain, wind, ice. The chippings made by these sculptors, which carve canyons and wear down mountains, are carried to the ocean by rivers and winds. Here they build up continental margins, with the left-overs accumulating in abyssal plains.

Much of the sediment comes to the sea through quasi-catastrophic events: floods, storms, earthquakes or — on a longer time-scale — ice advances. Other types of sediment arrive at the sea floor as a more or less continuous rain of particles: shells of plankton organisms, wind-borne dust, cosmic spherules. Through geologic time these gradually accumulating pelagic sediments built up a layer of a few hundred meters thickness, which forms a veneer on the oceanic crust. This veneer contains a detailed history of the evolution of ocean circulation and of pelagic organisms, for the last 100 to 150 million years.

Although exogenic processes tend to *level* the Earth, by erosion and deposition, they also can build mountains. The outstanding example is the Great Barrier Reef off eastern Australia, whose mountain tops rise thousands of meters above the Coral Sea. The mesa-like reef mountains are made from the calcium carbonate secreted by coralline algae, stony corals, molluscs and small unicellular organisms called foraminifera. The algae, of course, depend on sunlight. The corals and foraminifera contain unicellular algae within their bodies, in symbiosis: they too depend on sunlight for growth.

With this briefest of all introductions to the opposing effects of endogenic processes (which wrinkle Earth's surface) and exogenic processes (which mainly smooth it), let us now return to the nature of the grand morphology of the sea floor — and to this strange notion of sea floor spreading.

How did this idea arise in the first place?

1.4 Historical Aspects

Only some twenty years ago, it was still possible for geologists to think that the sediments on the deep sea floor might contain the entire Phanerozoic record and lead us back even into the Precambrian (see Appendix A3). Today, there would be few indeed who would harbor such fond hopes. The most ancient sediments recovered from the sea floor are about 150 million years old, which is less than 10% of the age accorded to fossil-bearing sedimentary deposits on land. Where then is the debris which must have washed into the deep sea, for the several billion years that continents existed?

14

According to the hypothesis of sea floor spreading, all sediments accumulating on the ocean bottom are swept toward the trenches, as on a conveyor belt (Fig. 1.5). Here some of the sediments are *subducted* into the mantle, others are scraped off against the inner wall of the trench. Thus the ocean floor is cleaned of sediments by constant renewal and destruction.

This remarkable idea did not exactly find a warm welcome when first proposed by H. H. Hess (1960), and by R. S. Dietz (1961). Somewhat similar ideas had been advanced earlier (Fig. 1.7) and had likewise been discounted as premature speculations. Gradually, however, as more facts became available to test the hypothesis, opposition weakened, and by 1970 there were only a very few defenders of tradition who challenged the concept of a moving sea floor.

The opposition to large-scale horizontal movement on Earth's surface had a long history. It all began with *Continental Drift*. Drastic changes in the distribution of continents and ocean basins had been proposed by the German geophysicist A. Wegener early in the century (1912). He envisioned granitic continents floating in basaltic mantle magma like icebergs in water. In fact, gravity measurements support this view, as does the gross nature of the hypsographic curve (Fig. 1.1).

Wegener also noted that many puzzling problems of paleoclimatology and paleobiogeography could be solved by assuming that the continents on either side of the Atlantic were once joined and drifted apart since. The "fit" of the continents, as in a jigsaw puzzle, was the symbol and chief argument of his concept. However, he also proposed that the continents plowed through the magma which carries them, an idea which proved to be wrong. In addition, he had a time table for the drifting of certain landmasses which was quite unrealistic. Skeptical geophysicists recognized these weaknesses in Wegener's proposal and argued forcibly for the rejection of continental drift. Thus, despite the support from geologists familiar with the incredible similarities of ancient rocks and fossils in South America and South Africa, Wegener's hypothesis did not find general acceptance.

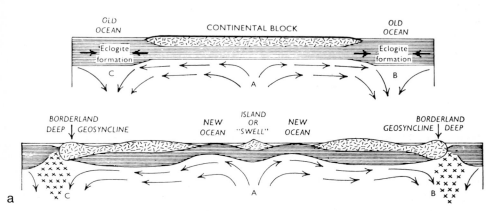

Fig. 1.7a–c. Models of sea floor spreading proposed to explain the origin of the Mid-Atlantic Ridge. **a** Hypothesis of A. Holmes (1929, Trans Geol Soc Glasgow 18: 559). Note the left-over continental piece in the center (which does not in fact exist).

15

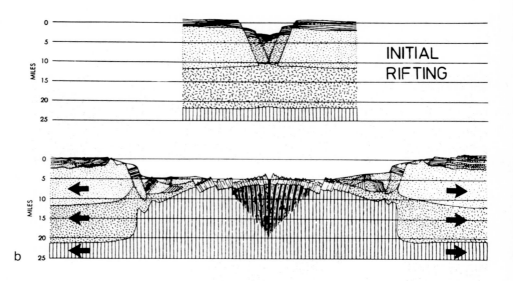

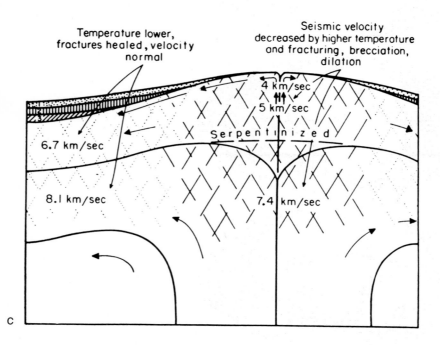

Fig. 1.7b. Hypothesis of B. C. Heezen (1960, Sci Am 203: 98). Note the exaggerated mobility of continental margins, and the postulated diffuse injection of mantle material in a broad ridge area. (In fact, injection occurs in a narrow zone.)

c Hypothesis of H. H. Hess (1962). Note narrow injection area, and stratigraphic onlap of deep sea sediments. This is the favored model, although serpentinization (a type of chemical alteration of basalt) was found to be much less important than visualized by Hess

16

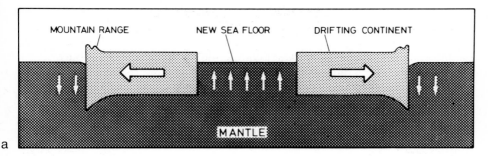

a

MOUNTAIN RANGE   NEW SEA FLOOR   DRIFTING CONTINENT

MANTLE

END OF PERMIAN

END OF JURASSIC

END OF CRETACEOUS

b

Fig. 1.8a, b. Diagrammatic representation (ours, not his) of the Hypothesis of Continental Drift of Alfred Wegener.
a Continental blocks made of light "sial" float iceberg-like in the heavier mantle "sima". As they drift apart, mountain ranges form at the bow, while new sea floor forms in between.
b The break-up of Pangaea, first envisaged by A. Wegener, in a modern reconstruction by R. S. Dietz and J. C. Holden (1970, J Geophys Res 75: 4939), simplified

17

In essence, evidence for continental drift was dismissed because the proposed mechanism was wrong. Only in the late 1950's through the work on geomagnetism and polar wandering by E. Irving and S. K. Runcorn, did it become possible again to talk about continental drift without being thought ignorant of physical principles. The compelling evidence, however, came from the magnetism of the sea floor itself, as we shall see.

1.5 Morphology of the Mid-Ocean Ridge

Ultimately it took geophysical evidence based on crustal magnetism to compel acceptance of the mobility of the ocean floor, and to turn the hypothesis of sea floor spreading into the ruling theory. The most obvious achievement of the new theory is the explanation of the origin of the Mid-Ocean Ridge, the central morphologic feature of the sea floor. The Mid-Ocean Mountain Range is more than 60,000 km long and takes up one third of the ocean floor, that is, about one fourth of the Earth's surface. In the Atlantic and along certain other portions, the crest is marked by a central rift, a 30- to 50-km-wide steep-walled valley 1 km deep, or more (see Figs. 1.2 and 1.3). The crestal morphology is usually very rugged and complicated, while the flanks tend to be smoothed by sediment (Fig. 1.9). The following is a brief account of how the theory of sea floor spreading explains the character of the Mid-Ocean Ridge.

The crest is characterized by shallow earthquakes (centers at less than about 60 km deep), by active volcanism, and by high heat flow values. The upwelling and spreading of mantle material pulls apart the crust, producing the central rift and generating the earthquakes. It also brings up heat from the Earth's interior. The *spreading rate,* that is, the rate at which sea floor on one side moves away from that on the other, is on the order of 1 to 10 cm per year. The hot mantle material filling the gap is less dense than old oceanic crust, because of thermal expansion. Away from the central Ridge it becomes denser as it cools, but on the whole, the

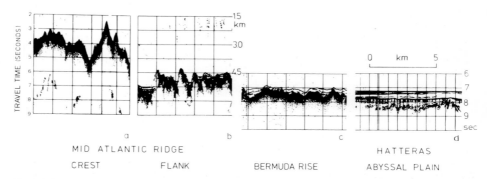

Fig 1.9. Portions of continuous seismic profiling records, from Mid-Atlantic Ridge *(left)* to Hatteras Abyssal Plain. (After T. L. Holcombe, 1977, Geo Journal 1 [6]: 31). Note the smoothing of the sea floor by a cover of sediment, with progressive age away from the central ridge area

18

new lithosphere and its sea floor float up higher out of the mantle and thus protrude, forming the Ridge.

Generally, the Ridge crest has an elevation (i. e., depth) of −2500 to −3000 m, all around the world. From this similarity of elevations, we may conclude that the upwelling material and its temperature are rather uniform.

With few exceptions the volcanic rocks forming the oceanic crust on the ridges are *olivine tholeiites*. They are dense, heavy silicate rocks rich in iron and magnesium, and belong to the basalts. Their minerals are essentially plagioclase feldspar, pyroxene, and olivine. Compared with the more familiar basalts on land they have low contents of potassium, titanium, and phosphorus. Also, they are depleted in those trace elements (rubidium, cesium, barium, lanthanium) which tend to be preferentially concentrated in the liquid phase during melting or during fractional crystallization. It is difficult, however, to deduce directly the composition of the mantle material from that of the tholeiite basalts. Many processes affect the magmas on their way up toward the sea floor. These processes include differentiation through partial melting; mixing of, and reactions between various types of melts; reactions with solution including seawater, and escape of gases (Appendix A7).

As the sea floor spreads away from the crest, the lithosphere cools and sinks, about 1000 m during the first 10 million years. The next 1000 m of sinking take about 26 million years (Fig. 1.10). It can be argued from physical principles that

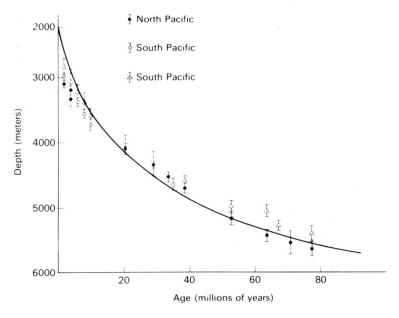

Fig. 1.10. Age–depth subsidence curve for the Pacific ocean floor. *Symbols* actual measurements. *Curve* calculated hypothetical profile for a cooling lithosphere 100 km thick. (After J. G. Sclater et al., 1971, J Geophys Res 76: 7888). Curves for other ocean regions are similar, regardless of spreading rate

19

the depth of the sea floor on the spreading flanks of the Ridge should be a simple function of age:

$$\text{depth below crest} = k \cdot \sqrt{\text{age}} \qquad (1.1)$$

From the relationship given above, we can evaluate k as follows:

$$k = \frac{1000}{\sqrt{10}} \quad \text{and} \quad k = \frac{2000}{\sqrt{10+26}}$$

which comes out as k≈320, when depth is in meters and age in million years. If this is correct, we can calculate the average age of the deep sea floor from its average depth (after correcting for sediment cover). For an average (corrected) basement depth of 5000 m (=2400 m below ridge crest) we obtain an age of 60 million years, which is indeed close to the average age of the sea floor.

During sinking of the sea floor, the rough topography which moves down the ridge flanks is gradually smoothed by the sediment cover (see Fig. 1.9). However, abyssal hills with a relief in the 50 to 1000 m range and slopes with 1° to 15° remain as expressions of the underlying basement morphology over large regions. Abyssal-hill morphology is the most common type of landscape on the face of Earth: in the Pacific Ocean about 80% of the sea floor belongs to this category.

1.6 Morphology of the Trenches

In general, trenches are found near the margins of ocean basins, notably the Pacific Basin. It is not obvious why there are not more mid-ocean trenches. To answer this question we need to learn more about processes within the mantle. First, some observations: trenches are roughly 100 km wide (in their shallower part) and from hundreds to thousands of kilometers long. For example, the Aleutian Trench is 2900 km long. The cross-section is usually V-shaped (Fig. 1.11), and the deepest part may be flat due to ponded sediment. Such sediments

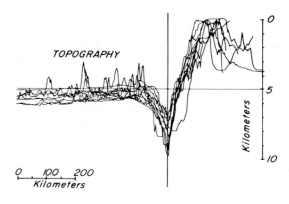

Fig. 1.11. Topographic profiles across trenches in various regions of the ocean. Land or island arcs to the *right*. (After M. Talwani, 1970, in The Sea 4 (1): 282)

20

generally show undisturbed horizontal layering — an observation which was sometimes used as an argument against subduction. The trench walls usually have slopes between 8° and 15°. However, steep sides (up to 45°) as well as steps also have been mapped. In cases outcrops of basalt have been observed by dredging and deep sea photography.

Trench depths listed vary (see R. L. Fisher, H. H. Hess, in The Sea v. 3, 1963; ref. p. 250). The greatest depths are in the western Pacific, in sediment-starved trenches off island arcs: Mariana Trench maximally 10,915 m; Tonga Trench, 10,800 m; Philippine Trench, 10,055 m; Japan Trench, 9,700 m; Kermadec Trench, 10,050 m. The values are not exact; they were determined from echo soundings which include corrections for effects of regional temperature and salinity distribution on the velocity of sound in the water. Any errors in the determinations, however, cannot mask the similarity in these depths. As in the similarity of ridge-crest elevations, this coherence points to the action of similar processes in each of the trenches in the western Pacific. Elsewhere trenches are shallower: Puerto Rico Trench about 8600 m; South Sandwich, 8260 m; Sunda 7135 m. The eastern Pacific is characterized by trenches directly adjacent to continents, without intervening island arcs. These trenches are filled with continental debris and this, presumably, is the reason they are shallower than their western counterparts (< 1000 m).

The ring of trenches girdling the Pacific is the site of most of the earthquakes on Earth: more than 80% of the shallow earthquakes (<60 km deep), 90% of the intermediate ones (60 to 300 km deep), and almost all of the deep quake centers (300 to 700 km) are concentrated here (Fig. 1.12).

The rest of the deep and intermediate quakes also occur largely in trenches; some occur in the Mediterranean, in Iran, and in central Asia at the northern boundary of the Himalayas. When the depths of the quake centers are plotted below the epicenters, they are seen to occur on planes which intersect the surface near the trenches and dip underneath the arcs or the continents at an angle of between 15°and 75°, to a depth of about 700 km. It is reasonable to conclude that the earthquakes are caused by friction on the upper surface of the downgoing slab of sea floor, and that this friction ceases when the temperature becomes high enough to cause flow (Fig. 1.13).

The "ring of fire" around the Pacific is closely associated with the trenches: volcanoes sit on top of the dipping earthquake planes, that is, the downgoing lithosphere. Of 800 active volcanoes, 75% are in the "ring". Where the lithosphere descends before reaching the continents, the volcanoes form island arcs; where it descends under a continent (South America) mountain ranges are formed. The magmas produced by partial melting of the downgoing slab mix with overlying materials on their way up and form characteristic volcanic rock types, the *andesites* — named after the Andes Mountains (Appendix A6).

According to the theory of sea floor spreading, the trenches are produced by the subduction of sea floor. The descending lithosphere, some 100 km thick, finally disappears into the asthenosphere, that is, the "soft" part of the upper mantle. The descending lithosphere offers up materials to the continent, by the scraping off of sediment, and by partial melting. These materials contribute to continental accretion, that is to the growth of continents.

21

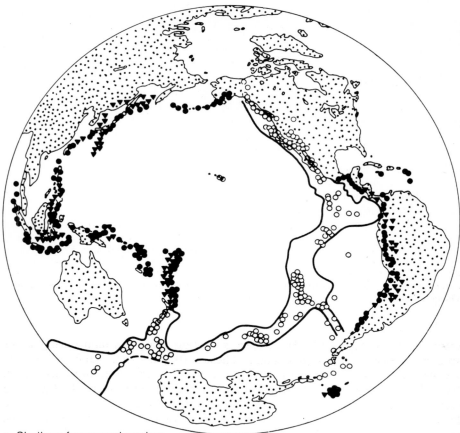

o Shallow-focus earthquakes
● Intermediate-focus earthquakes
▼ Deep-focus earthquakes

Fig. 1.12. Earthquake belts in the Pacific. (After R. W. Girdler, 1964, Astron Soc Geophys J 8: 537). Deep and intermediate earthquakes are restricted to trench regions

Continental accretion is one way in which the endogenic forces oppose the wearing down of the continents by exogenic agents. Thus, the continued existence of continents which rise high above the sea floor, is intimately tied to the processes of sea floor spreading.

1.7 Fracture Zones and Plate Tectonics

So far we have glossed over the fact that the Ridge Crest is not continuous but occurs in more or less straight portions which are offset from each other. The consequence of such offset is that a lateral fault must form at the two ends of each

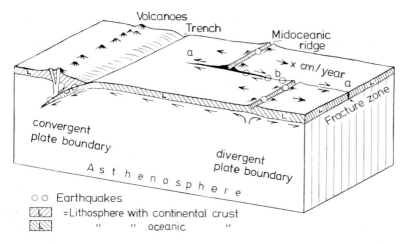

Fig. 1.13. Relationship of plate motions to plate boundaries and earthquakes. *Divergent plate boundary* Mid-Ocean Ridge, shallow earthquakes. *Convergent plate boundary* Trench, shallow, intermediate, and deep earthquakes. Lateral boundary: *fracture zone,* shallow earthquakes on active part only. Horizontal motions are of the order of 1 to 10 cm/yr. (Based on a diagram by B. Isacks, J. Oliver, and L. R. Sykes, 1968, J Geophys Res 73: 5855)

crestal portion (Fig. 1.13). Since there is motion along this fault during active spreading, there are earthquakes on it. These earthquakes are shallow and define the *active* part of the *fracture zone,* that is, the *ridge-ridge transform fault.* Beyond this active part, the fracture zone is the frozen trace of the fault; the scarps subside as the sea floor ages on both sides of the zone. These extensive linear zones have an unusually irregular topography with large sea mounts, steep-sided or asymmetrical ridges, troughs, or escarpments (Fig. 1.3).

Some fracture zones connect the end of a ridge crest portion to a trench. These zones are seismically active and constitute the third type of boundary that defines a lithospheric slab or *plate.* The other two, of course, are *spreading center* and *trench.* The fact that these boundaries form "plates" was first pointed out by J. T. Wilson, in 1965. On the basis of earthquake distributions and first motion studies (that is, observing which way the ground moves upon initiation of a quake), it is possible to outline a number of large lithospheric slabs, dividing the surface of the globe. Each of the plates has its own particular motion, which can be read from the magnetism of the sea floor, as we shall see. The quantitative development of these concepts was initiated in the late sixties by D. P. McKenzie and R. L. Parker; W. J. Morgan; X. LePichon; and by B. Isacks, I. Oliver and L. R. Sykes.

The motions are generally uniform and do not result in deformation of the plates; hence they can be described as rotations on a sphere, according to a theorem of the famous mathematician, Leonhard Euler (1707–1783). The fracture zones provide traces for the latitudinal circles around the pole of rotation (which need not coincide with that of the rotation of Earth). Thus, the pole of rotation can be determined for each plate. Geometry requires that spreading rates must increase away from the pole of rotation for separating plates, and this is indeed observed. It will be noted in Fig. 1.14 that a plate can contain both oceanic and continental lithosphere. In fact, the continents share the motions of the mobile ocean floor. Thus, the continents do drift, as Wegener had supposed, but not by plowing through the mantle magma.

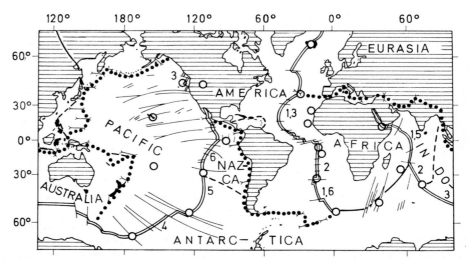

Fig. 1.14. Major lithospheric plates of Earth, as proposed by W. J. Morgan (1968, J Geophys Res 73: 1959). Divergent boundaries: *double lines* (*numbers* indicate spreading rate in cm/yr). Convergent boundaries: *solid dots* (trenches and alpine mountain chains). Plumes or "hot spot" volcanism: *open circles*

1.8 Seamounts, Island Chains, and Hot Spots

With few exceptions, oceanic islands are made of volcanic rock, with or without a crown of reef carbonate. A crown of reef carbonate, of course, can only be precipitated in shallow water because it depends on algal growth, as mentioned. Thus, if a seamount is found with a top of reef carbonate, and deeply submerged below the present sea level, it must have sunk. Such seamounts are in fact common in the western Pacific.

It has been said that the discovery of flat-topped seamounts held the key to the new understanding of the origin of ocean basins. Flat-topped seamounts were first discovered in the forties by H. H. Hess (Fig. P.1) in the central Pacific. Hess proposed that these table mounts, the *guyots,* as he named them, had formed as volcanic islands, were truncated by wave erosion, and then sank to their present depths. He also initially thought they might be of Precambrian age, with lots of time available for subsidence. However, no rocks older than Cretaceous were ever dredged from the guyots.

24

In essence, Hess' hypothesis of guyot formation was an extrapolation of Charles Darwin's hypothesis of atoll formation (see Chap. 7). The idea of seamount subsidence was easily reconciled with Hess' later concept of sea-floor spreading (Fig. 1.7c). Thus, he discovered both a major problem — the origin of guyots — and its solution: sea-floor spreading. Seamounts in general have elevations of more than 1000 m and typical slopes of 5° to 15°. The Pacific has about 10,000 of such seamounts.

There are a number of striking instances where seamounts — either flat-topped or not — occur in linear chains. The Hawaiian chain is a prime example. How are such chains generated? One possible explanation which has been proposed is to assume that the volcanoes making the chain lie on a long zone of weakness in the crust, a deep fracture, along which magma can rise to form volcanic islands. Yet, at least in the case of the Hawaiian Islands, there is clearly a progression from high, large islands with active volcanoes at the tip of the line, to sunken islands with extinct volcanoes at the end (Fig. 1.15). It certainly looks as though the big islands are young and the sunken ones old. Dating of the rocks, by radioactivity, confirms this impression. Thus, in the fracture hypothesis, we have to postulate a propagating crack, which opens at one end and closes at the other.

A more satisfactory explanation of the origin of island chains was given by J. T. Wilson, (in 1965) and by W. J. Morgan (in 1972). They proposed a stationary source of hot magma, deep in the mantle, over which the lithosphere rides. Volcanoes build up on top of the crust over the "hot spot". As the plate moves, a trail of extinct volcanoes forms behind the active tip of the line (Fig. 1.15).

This trail, then, indicates the direction of movement of the lithospheric plate, with respect to the (more or less stationary) mantle source. Changes in direction of the trail, as between Hawaiian islands and Emperor seamounts, would indicate changes in the direction of plate motion.

1.9 "Proof" for Sea Floor Spreading: the Magnetic Stripes

So far, we have used the concept of sea floor spreading for explaining a number of important morphological features of ocean basins. What proof do we have that the theory is sound? This question has been raised well into the seventies and is still being posed by some.

What is "proof" in the geologic sciences? Can we prove that a fossil was once part of a living organism? Can we prove that the Earth has an age of 4.6 billion years? Can we prove that large continental glaciers once covered vast areas of North America and northern Europe?

It all depends on what one is willing to accept as proof. Certainly all the above questions were once vehemently answered with NO by experts, and would be answered with disbelief today — disbelief that anyone would ask such a silly question.

What then about proof for sea floor spreading? Since 1968, the year several major articles appeared on the subject, "sea floor spreading" has joined the various other propositions about the Earth which are accepted as fact. We can

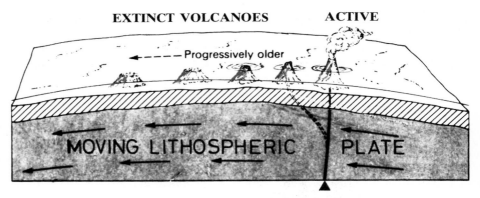

EXTINCT VOLCANOES   ACTIVE

— Progressively older —

MOVING LITHOSPHERIC PLATE

STATIONARY MANTLE PLUME
(HOT SPOT)

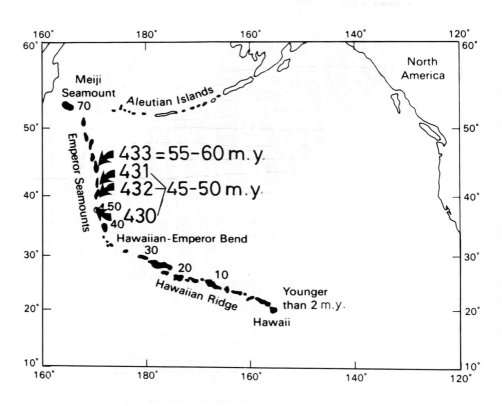

Fig. 1.15. Origin of Hawaiian islands, submerged coral banks, and Emperor Seamounts, according to the "hot spot" hypothesis as envisaged by J. T. Wilson (1963, Can J Phys 41: 863) and W. J. Morgan (1971, Nature (London) 230: 42). *Upper* sketch of hypothesis. *Lower* ages in million years along Hawaiian Ridge (K-Ar determinations, summarized by S. Uyeda [1978] and biostratigraphic ages of basal sediment in Glomar Challenger Sites, 430–433 Leg 55, 1977)

now cease to quote dissenting voices, just as we no longer bring up the "plays of Nature" when explaining fossils, or Lord Kelvin's calculations about the age of the Sun, or the "Great Flood" when explaining the Ice Age deposits.

Why can we be so confident that the sea floor spreading story is correct?

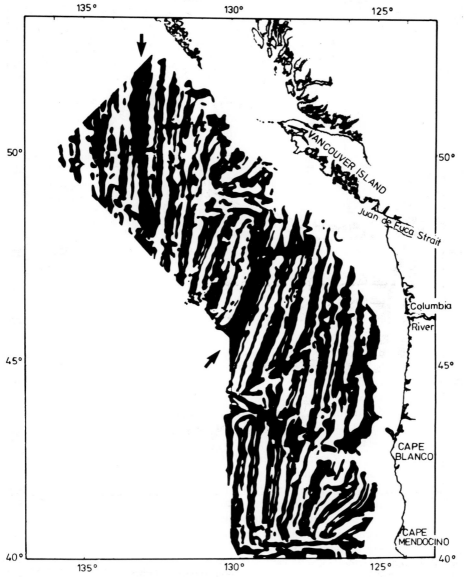

a

Fig. 1.16. Magnetic lineations on the sea floor. **a** Anomaly patterns of A. D. Raff and R. G. Mason (1961, Geol Soc Am Bull 72: 1267). These anomalies remained unexplained for several years. They are now recognized as being generated at the "spreading centers" *(arrows)*

27

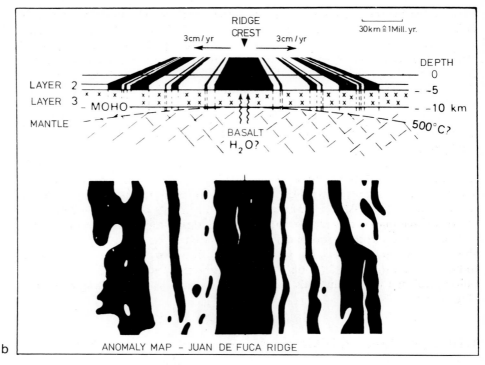

Fig. 1.16b. Mechanism of generating magnetic anomalies, as proposed by F. J. Vine and D. H. Matthews (1963, Nature [London] 199: 947). From F. J. Vine in R. A. Phinney, 1968, The History of the Earth's Crust, Princeton Univ. Press pp 73–89)

Our confidence derives from the global pattern of magnetic anomalies on the sea floor (Fig. 1.16). For each of the observations on morphology, heat flow, seismic activity, etc., which are so nicely explained by sea-floor spreading, it is possible to conceive of some other way to produce the phenomenon. However, for the magnetic anomalies, no reasonable alternative to sea-floor spreading has ever been proposed — and not for lack of trying.

The patterns, the "magnetic stripes", were first discovered by geophysicists at Scripps Institution of Oceanography (R. G. Mason, A. D. Raff, V. Vacquier). However, their origin remained a complete mystery for several years. One of the problems was that the area for which they had been mapped is tectonically complicated, and the symmetry of the patterns about the Ridge, which holds the key to the explanation, is not obvious at all.

The first successful attempt to account for the "stripes" was by F. J. Vine (then a graduate student at Cambridge University) and D. H. Matthews (his advisor), in 1963. Their suggestion was strikingly simple: put together the ideas on sea floor spreading of H. H. Hess and of R. S. Dietz and combine them with the evidence for periodic reversals in the Earth's magnetic field, as presented by A. Cox and co-workers (in 1963). The newly upwelled, hot material at the Ridge

28

Crest (or within the central rift) is magnetized upon cooling below 525° (the Curie Point), in accordance with the prevailing magnetic field. If this field reverses periodically and, they said, "if spreading of the ocean floor occurs, blocks of alternately normal and reversely magnetized material would drift away from the center of the ridge and parallel to the crest of it." Here, in a nutshell, was the key to proving the reality of sea-floor spreading (Fig. 1.17). The sea floor could be seen to act as a tape recorder of the Earth's magnetic field.

Eventually, the magnetic anomaly sequence on each side of the ridge, in every major ocean basin, was found to be exactly the same as in the lava flows studied (and dated) on land.

The tape recorder is perhaps not of the finest quality, but it works remarkably well (Fig. 1.17a). With a time scale at hand (Fig. 1.17c), we can now read off the spreading rate, simply by matching the sea floor anomalies to the magnetic reversal scale. Thus, we can make an age map of the sea floor (Fig. 1.18).

If the age map is correct, we should then find that the oldest sediment lying on the basaltic substrate shows the same age progression. The deep sea drilling ship Glomar Challenger set out in 1968 to test this prediction. It was first found to be correct during Leg 3 (1968/69). Since then, most magnetic and micropaleontological age determinations have coincided with an astonishing exactness (Fig. 1.19).

1.10 Open Tasks and Questions

Many of the open questions concern refinement of existing concepts. Improving the reversal time scale ist one of the most important tasks, because the scale forms the basis for work on rates of motion and sedimentation. Others are more ambitious: Did plate motions change drastically at certain times? What is the origin of magnetically quiet zones with only weak lineations, which occur at Atlantic margins? What is the origin of unexpected magnetic anomalies, that is, those which do not follow the established sequence?

Yet other questions, more fundamental, are concerned with mechanisms. What does convection in the mantle actually look like? What is its relationship to plate motions? What controls the rates of spreading? Why is spreading slow in the Atlantic and fast in the Pacific?

For the marine geologist interested in reconstructing the history of the oceans, reliable paleogeographic maps showing the distribution of land and water are perhaps the most pressing need. E. C. Bullard et al., in 1965, demonstrated how to reassemble correctly the drifting continents and continental fragments on a sphere (Fig. 1.20). Bullard's co-workers (A. Smith and colleagues) have greatly extended this type of work and have recently supplied a series of maps with ancient positions of continental masses. These, of course, are extremely useful for historical geology, both continental and marine. However, we know that mountain building and other processes active at the transitions between oceanic and continental crust have changed significant details of the continental configurations. Some of these changes are crucial in deciding, for example,

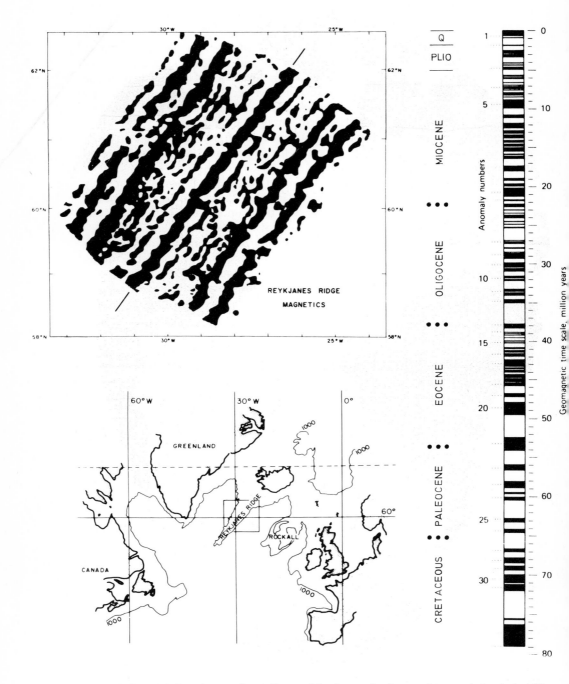

Fig. 1.17. Magnetic lineations on the sea floor and the time-scale of magnetic reversals, for the last 80 million years. Pattern on Reykjanes Ridge from J. R. Heirtzler et al. (1966, Deep-Sea Res 13: 427). Time-scale after Heirtzler et al. (1968, J Geophys Res 73: 2119). Biostratigraphic boundaries modified

30

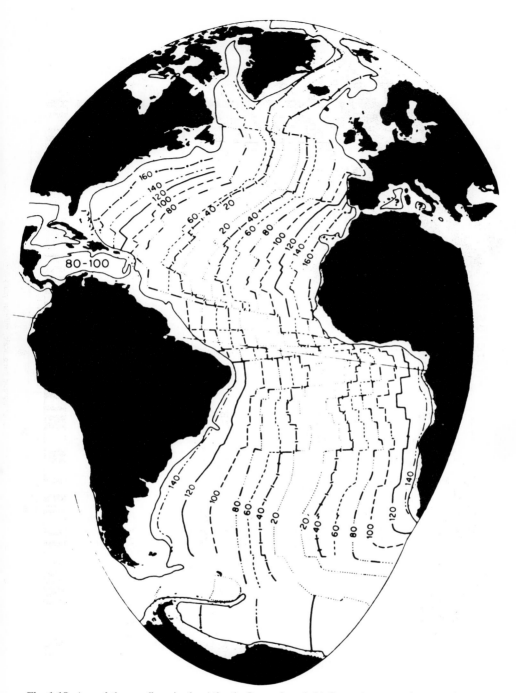

Fig. 1.18. Age of the sea floor in the Atlantic Ocean, based chiefly on the magnetic reversal scale. (From W. H. Berger, E. L. Winterer, 1974, Int Assoc Sediment Spec Publ 1 : 11). The gradual increase in size of the Atlantic is evident. It grows at the expense of the Pacific (whose age distributions are less well established)

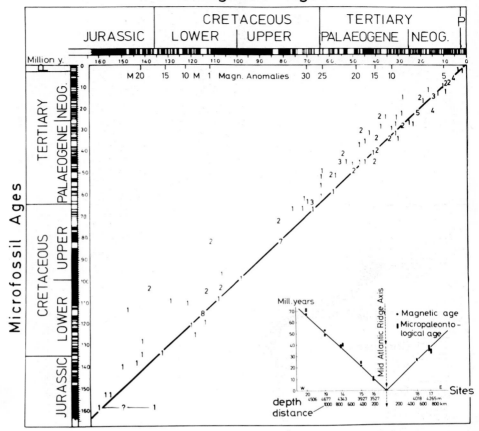

Fig. 1.19. Comparison of magnetic anomaly ages ("basement ages") and biostratigraphic ages for oldest sediment, Glomar Challenger Sites 1–417 (1968–1976). Compiled by Dr. M. Sarnthein, Kiel. Note that most of the sediment ages are slightly younger than the paleomagnetic basement ages as is expected. *Insert* age–distance plot of DSDP Leg 3 data, on a profile across the Mid-Atlantic Ridge off Brazil at 30° S. These data (published by A. E. Maxwell et al., 1970, Science 168: 1047) first demonstrated the agreement between paleomagnetic dating based on the hypothesis of sea floor spreading, and the dates derived from biostratigraphy

whether there was a connection between one ocean basin and another. It will take many years of compilation, fieldwork, and detailed reassembly just to provide the kind of paleogeographic base maps necessary for explaining the distribution of ancient fossils, for example.

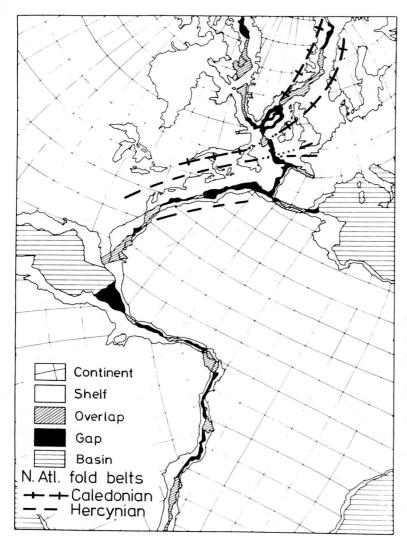

Fig. 1.20. The fit of the continents bordering the Atlantic, as proposed by E. C. Bullard et al. (in Blackett et al. 1965). The fit is based on matching the 500 fathom (900 m) contour. Note the Niger Delta overlap, which is expected, and the Bahamas overlap, which is unexplained. Note also the position of Gibraltar relative to Africa, and the "closure" of the Bay of Biscay. The alignment of the Paleozoic fold belts in North America and Europe (here added) is remarkably good

2 Origin and Morphology of Ocean Margins

2.1 General Features of Continental Margins

The continents are very old: up to thousands of millions of years. Ultimately they are the product of selective accumulation of low density mantle material. Because of this low density they float in the mantle. The ocean floor, on the other hand, is geologically young, as we have seen. The basaltic rock which forms the ocean floor basement is rather close in composition to the mantle rock it came from. It is slightly heavier than continental rock (largely due to its high iron content). The light-weight continental mass protrudes above the surrounding sea floor (Fig. 2.1a). Thick sediment piles accumulate at the boundary between continent and ocean, which build the actual margin (Fig. 2.1c). These sediments may be well-layered or strongly deformed, depending on the tectonic forces active at the margin.

The ocean margins, that is, the regions of transition between continent and deep ocean, differ greatly in their characteristics, depending on whether they occur in mid-plate areas (on the continent's trailing edge), or on the collision edge of a continent, or along a shear zone. The one thing most of the ocean margins have in common is the occurrence of large masses of sediment. Ocean margins are generally referred to as *continental margins* — a reflection of our landlubber point of view.

The importance of the continental margins in the overall geography of the ocean floor can best be illustrated with a few statistics (Table 2.1). The various numbers reflect in essence the efficiency of the exogenic and endogenic processes which provide the balance between the extent of continents and of ocean basins. This balance is produced by erosion of highlands, deposition around the continents, and the mountain-building processes briefly alluded to earlier.

From Table 2.1 we see that 70% of the *land area* (20.7/29.0) is within 1000 m above sea level. Continents wear down toward sea level because it is the baseline of erosion. Sea level is also the top level of deposition. Hence the sediments deposited offshore tend to build up to sea level. The great plains of the lower Mississippi Basin, and the entire Gulf Coast are prime examples of this tendency for large continental areas to be near sea level. These areas are underlain by sediments deposited close to sea level during times when the sea invaded the continent.

Large parts of the low-lying portions of continents are covered with marine deposits. Actually, such areas are part of the *shelf* of a continent, a part which is frequently submerged in the course of geologic history, but which happens to be exposed right now. At the present time in our geologic period, an unusually large

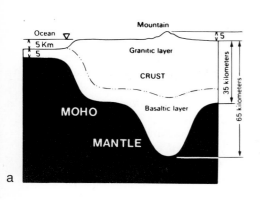

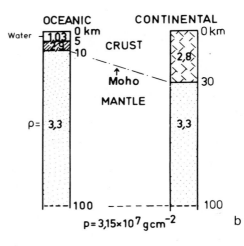

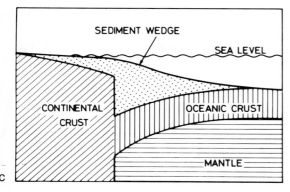

Fig. 2.1a–c. Isostatic block model for continent-ocean-transition. **a** Schematic cross-section through continent "floating" on mantle (Uyeda, 1978). **b** Density profiles. **c** Sketch of general nature of continental margin

Table 2.1 Statistics on continental margins (after H. W. Menard and S. M. Smith, 1966, J Geophys Res 71 p 4305, and other sources)

World ocean without adjacent seas	Shelf			Continental slope			C. Rise	Trenches
	Area (0–200 m) 10^6 km²	Average width (km)	Average slope	Area (10^6 km²)	Average width (km)	Average slope	Area (10^6 km²)	Area (10^6 km²)
Atlantic	6.080			6.578			5.381	0.447
(% of its area)	(7.0%)	115	0°28'	(7.6%)	260	1°19'	(6.2%)	(0.5%)
Indian Ocean	2.622			3.475			4.212	0.256
	(3.6%)	91	0°23'	(4.7%)	182	1°35'	(5.7%)	(0.3%)
Pacific	2.712			8.587			2.690	4.757
	(1.6%)	52	0°49'	(5.2%)	139	3°13'	(1.6%)	(2.9%)

35

proportion of this shelf is exposed. If we put the outer end of the shelf at 150 m water depth (for convenience — it varies greatly), we see that about 5% of the Earth's surface consists of submerged shelves (Table 2.1), that is, between 7% and 8% of the ocean floor, or nearly the area of Africa.

The schematic drawing in Fig. 2.2a shows the relationships between shelf, continental slope, and continental rise, and introduces commonly used terms associated with continental margin environments. The physiographic diagram of the margin off the East Coast illustrates the various morphologic provinces (Fig. 2.2b).

The terms *pelagic* and *neritic* refer to marine organisms as well as sediments, and mean *open ocean* and *coastal*, respectively. *Littoral*, etc. down to *hadal*, refer to water depths. *Littoral* is the same as *intertidal*, meaning between tides. *Supralittoral* refers to the spray zone, *sublittoral* to offshore from the tidal area.

2.2 Margins Are Sediment Traps

The continental margins are the dumping site for the debris coming from the continents, the *terrigenous* sediments. The margins are also the most fertile parts of the ocean, where productivity is high. Thus, much organic matter becomes buried within the continental debris. If conditions are right, over millions of years, this organic material can eventually develop into petroleum. This happened, of course, in the Gulf Coast area where oil is found buried under immense masses of sediment (see Chap. 10).

Table 2.2 Land and ocean areas and drainage (after H. W. Menard and S. M. Smith, 1966, J Geophys Res 71 p 4305, and other sources)

	Area (10^6 km^2)	% of Earth surface	Drained land area[b] (10^6 km^2)	Ocean area / land area	Average depth of water (m)
Asia	44.8	8.7			
Europe	10.4	2.1			
Africa	30.6	6.0			
North America	22.0	4.3			
South America	17.9	3.5			
Antarctic	15.6	3.1			
Australia	7.8	1.5			
Pacific	181.3[a] (166.2)	35.4	18	10:1	4.0 (4.2)
Atlantic	106.6[a] (86.6)	20.8	67	1,6:1	3.3 (3.8)
Indian Ocn	74.1[a] (73.4)	14.5	17	4,3:1	3.9 (3.9)

[a] Including adjacent seas (Black sea, Mediterranean and Arctic included with Atlantic). Numbers in parentheses: without adjacent seas.
[b] Excluding areas with interior drainage, and Antarctic

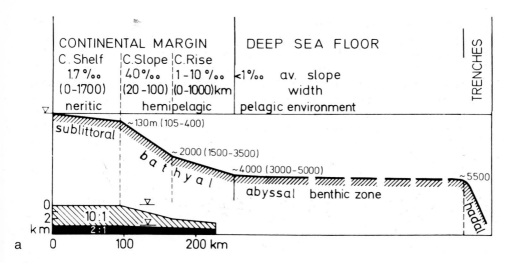

Fig. 2.2a, b. Depth zones of the sea floor. **a** Diagram defining the most common terms used in connection with sea-floor depth and distance from land. Profiles are usually strongly exaggerated; in reality slopes are gentle. The inset at the bottom shows this for a line off North West Africa. **b** Physiographic diagram of the continental margin of NE America (after B. C. Heezen et al., 1959, Geol Soc Am Spec Pap 65). *1* shelf; *2* continental slope; *3* continental rise; *4* abyssal plain; *5* submarine canyon

37

Which margins are likely to have thick sediment wedges? It is reasonable to expect that an ocean basin draining a large land area, per unit area of ocean (Table 2.2), is going to have a thick accumulation of sediment. Indeed, the margins of the Atlantic Ocean have very thick wedges of sediment, up to 10 km and more. The Atlantic also has the largest proportions of slope and continental rise areas of the major ocean basins (Fig. 2.3). The reason for this is not only sediment supply, but also the fact that the Atlantic margins are old "trailing edges" and have not been disturbed by tectonic processes, other than sinking, for a long time.

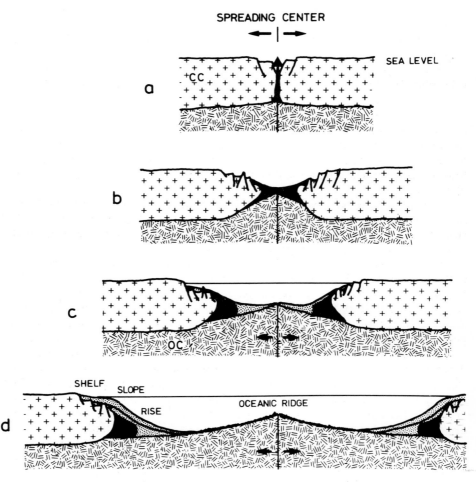

Fig. 2.3a–d. Evolution of Atlantic-type continental margins. Uplift of Earth mantle material **a** expands the continental crust *(CC)* causing graben structures. Volcanism is common at this stage. The continental crust thins, subsides, and **b** splits apart. Coarse terrigenous sediments, volcanogenic deposits (and salt in some cases) accumulate. Rifting is followed by drifting, with further subsidence of continental margins. Mantle material forms new oceanic crust *(DC)* as shown in **c**. This stage resembles modern Red Sea conditions. **d** Sea floor spreading widens newly formed oceanic crust area. Sediments cover older parts of sea floor, and build up margin

2.3 Atlantic-Type (=Passive) Margins

Continental margins differ greatly, depending on their origin. As early as 1883, E. Suess (1831–1914) coined the terms *Atlantic margins* and *Pacific margins* to emphasize the major differences. Essentially, Atlantic-type margins are steadily sinking regions, accumulating thick sequences of sediment in layer-cake fashion. In contrast, Pacific-type margins are, on the whole, rising, and are associated with volcanism, folding, faulting, and other mountain-building processes. Atlantic-type margins also are called "passive", and Pacific-type margins "active", because of the difference in tectonic style.

The origin of the margins must be understood in the context of sea floor spreading. In the Atlantic, the continental margins originated through a tearing apart of an ancient continent, along a line of weakness or great stress, and subsequently evolved through sinking and loading with sediment (Fig. 2.3).

The development can be illustrated taking the Red Sea as a model. Here, mantle material pushes up and tears the Arabian Peninsula from Africa. The edges of the opposing continental margins tend to break off and sink, as long narrow blocks. The mechanisms of the sinking are not fully understood. It is clear, however, that as the sea floor cools, it sinks, and thus the margin sitting next to it loses support. Also, gravitational sliding of the unsupported margin edges may play a role.

Reefs can grow on these sinking blocks, building up a carbonate shelf, and further depressing the crust with their weight. If the Red Sea were only slightly less open, salt deposits would form — indeed there is evidence from thick evaporite deposits that this happened in the past.

Thus, the receding margins sink and a rampart of reef carbonate may build up, and salt deposits also may form in the early phase of rifting.

Ancient salt deposits and reef ramparts are just what we see along many of the Atlantic margins (Fig. 2.4a). Salt deposits, of course, also are well known from the Gulf of Mexico, where they push up as salt domes (diapirs), providing a path for petroleum migration (see Chap. 10). In the Atlantic proper, good evidence for salt deposits exists, especially off Angola. The salt in the South Atlantic was laid down, presumably, when this ocean was narrow, was closed to the north, and had restricted exchange to the south due to the Walvis Ridge-Rio Grande Barrier (now near 30° S). Large petroleum reserves may be associated with these salt deposits, because the South Atlantic also was the site for deposition of organic-rich sediments, during a long period in the middle Cretaceous.

The kind of material accumulating on sinking continental margins depends on the geologic setting of the region. In the tropics, and where no large rivers bring sediment or fresh water, reef carbonates can grow. Elsewhere, mixtures of lagoonal and riverine sediments may gradually be buried by offshore deposits — mainly hemipelagic mud, rich in the shells of *planktonic* (floating) and *benthic* (bottom-living) organisms. In places, the sediments can become extraordinarily thick: 10 to 15 km of sediment are reported from off the Niger, the Mississippi, and from other large deltas.

The end-result of rifting, then, is continental margins consisting of thick

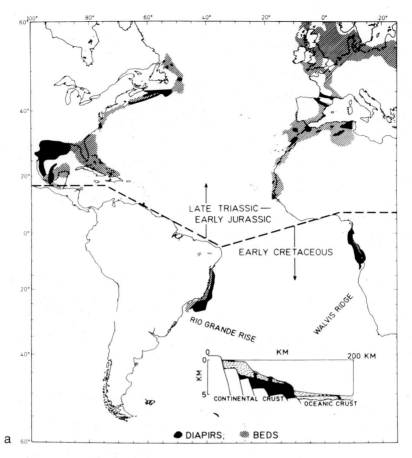

LATE TRIASSIC —
EARLY JURASSIC

EARLY CRETACEOUS

RIO GRANDE RISE

WALVIS RIDGE

KM   200 KM

CONTINENTAL CRUST   OCEANIC CRUST

● DIAPIRS;   ▨ BEDS

a

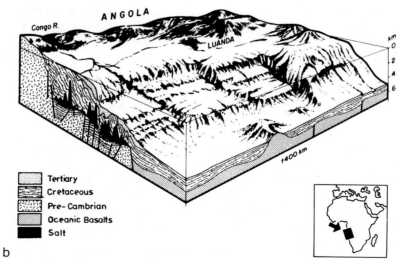

ANGOLA

Congo R.

LUANDA

km
0
2
4
6

1400 km

Tertiary
Cretaceous
Pre-Cambrian
Oceanic Basalts
Salt

b

40

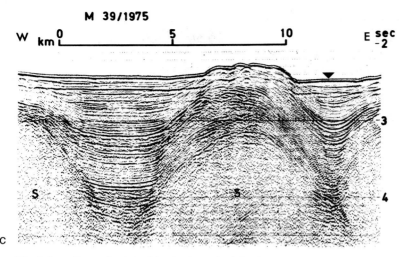

M 39/1975

W km 0   5   10   E sec
                                        -2

S   S

c

Fig. 2.4a–c. Evaporite deposition in the early Atlantic.
a Geographic distribution of Mesozoic evaporites (K. O. Emery, 1977, AAPG Continuing Education Course Notes Ser 5: B-1).
b Relationship of salt diapirs to margin structure off Angola (SW Africa). The Aptian salt is underlain by nonmarine clastic deposits which fill graben-like depressions within pre-Cambrian basement (R. H. Beck and P. Lehner, 1974, AAPG Bull. 58, 376).
c Salt diapir structures *(S)* as seen on air gun profile of Meteor Cruise 39, off Morocco (near 30° N). Water depth at triangle is approximately 1800 m. (From E. Seibold et al., 1976)

sediment stacks piled both on the sinking blocks of a continental edge, and on oceanic basement adjacent to the continental edge (Fig. 2.5).

We can generalize these conclusions to all margins which originated by rifting and are riding passively on the moving plate (hence *passive* margins). Besides the Atlantic margins proper, there are the East African margin, the margins of India, much of the margin of Australia and practically all of Antarctica. In the Antarctic, of course, special conditions prevail with respect to erosion and deposition, at least since the formation of ice sheets. Thick ice sheets have been present there since the late Tertiary and possibly earlier.

2.4 Unsolved Questions in the Study of Passive Margins

The unraveling of the exact origin and evolution of each stretch of passive margin poses its own problems. The commonly used analogy of the evolution of rifting — from the East African Rift to the Red Sea, to the Gulf of California, and finally the Atlantic — provides guidance as to which processes may be at work. Was there uplift before the rifting, and erosion of a crustal bulge at the site of future rifting? How wide was the original rift valley? How does the sinking of the outer parts of the continental edge affect landward crustal blocks? What are the rates of uplift and subsidence, and of erosion and sedimentation in time and space? With

41

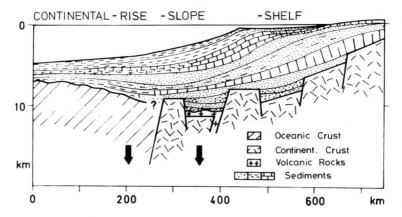

CONTINENTAL - RISE - SLOPE   - SHELF

Fig. 2.5. Sketch of an Atlantic-type margin in cross-section. The edge of the continental crust sinks because it rides on top of a cooling (and sinking) lithosphere. The sediment load produces additional sinking

regard to history of subsidence, what ist the relative role of "floating in the mantle" (isostatic equilibrium) of the blocks, *versus* gravitational sliding? What are the forces provoking the *uplift* of certain parts of the continental margin, and the formation of long deep-seated *barrier ridges* along some margins? What is the significance of the lack of sediments of a certain age, in many margins? Was the lack caused by erosion? By nondeposition? By huge landslides?

One question of fundamental interest is whether and where the thick sediment stacks piling up on the passive margins will eventually be found in the geologic record, on land. After all, the Atlantic cannot just go on rifting apart — sooner or later it runs out of space. One suggestion by J. T. Wilson is that a proto-Atlantic once was formed by rifting, *and then closed again,* running the previously passive margins into each other. The presumed product of this process: the chain of mountains from Norway through Scotland through Newfoundland and to the Appalachians. Check their positions on the "Bullard Fit" (Fig. 1.20) — Wilson's suggestion makes good sense.

If Wilson's hypothesis is correct, the passive margins would have turned into active margins when colliding with the trench that must have been there to make the proto-Atlantic disappear. What does such a collision margin look like? Would we be able to go to the mountains to check for the signs of collision?

To answer this question we must study the Pacific-type margins, that is, the collision margins.

2.5 Pacific-Type (= Active) Margins

We have previously alluded to the collision of a continent with a trench, focusing on the evidence for subduction (Sect. 1.6). Actually, there are two types of collision margins which we need to consider: those produced by continent–ocean

42

collision, as at the Peru-Chile Trench, and those where the subduction takes place along island arcs, as along the Marianas, for example (Fig. 2.6).

Perhaps the most important characteristics of collision margins are the folding and shearing of sediments, and especially the addition of volcanic and plutonic

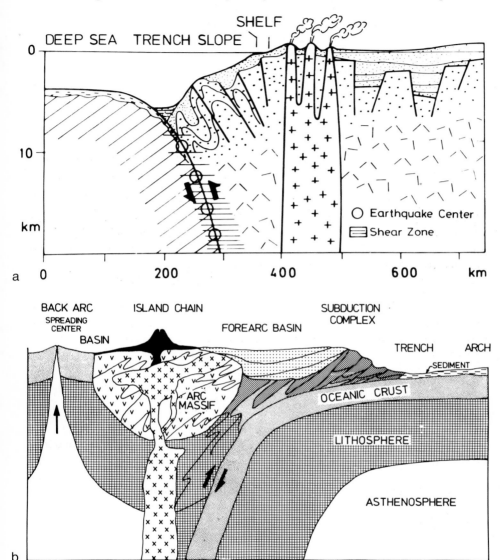

Fig. 2.6a, b. Sketch of collision margins, in profile. (Not to scale)
a Peru-type collision (ocean–continent). Slope sediments are being tectonically deformed. Igneous activity including volcanism derives from melts generated within the subduction zone.
b Island-arc situation (ocean–ocean). Volcanic islands build up over the subduction zone. Sources: see J. R. Curray, D. G. Moore, in C. A. Burk and C. L. Drake (1974) (ref. p. 250); and D. R. Seely, W. R. Dickinson (1977) Amer Assoc Petrol Geol Continuing Educ Notes Ser 5

material, from the active vents sitting on top of the down-going lithosphere. Also, the fractionation processes associated with partial melting on the descending slab and with hydrothermal reactions can lead to enrichment of melts with heavy metals — and hence to the formation of ore deposits, as in the Andes.

The types of rocks which characterize the continental margins next to subduction zones are extremely varied, which comes as no surprise. The incoming lithosphere brings an assortment of basaltic rocks, serpentinite, gabbro, peridotite which were derived from the mantle and altered by hydrothermal reactions under various conditions of pressure and temperature. In addition, various kinds of pelagic sediments may be added — deep sea clay, shell carbonates, biogenous silica. When such rock assemblages are found on land, they are referred to as *ophiolites* and are mapped in the hope of finding ancient subduction zones.

The steep slopes leading into the trench are favorable for large-scale gravitational transport of rock masses from the land side into the subduction zone. The jumbled masses *(mélange)* thus generated are then sheared, and metamorphosed (i. e., baked and cooked) under pressure (but at relatively low temperatures). Blue schists, green schists, and subsequently amphibolites can form under these conditions.

In classic geologic literature, the sediments of trailing edges are known as *miogeosynclinal,* and those of collision edges as *eugeosynclinal.* The *geosyncline* part of the terms, of course, stems from the observation that the Earth's crust must have subsided in order to accumulate the thick masses of sediment found in the mountains.

2.6 Shear Margins and Complex Margins

Passive or trailing continental margins normally are parallel to mid-oceanic ridges, as clearly seen in the North and South Atlantic. What about the nearly east-west-running margins of North Brazil and the African Guinea Coast? They are parallel to the many fracture zones near the Equator there and represent a third type, shear margins, with narrow shelves.

Not all margins can be readily classified. Some of these rim the marginal basins behind island arcs and elsewhere in adjacent seas. Some have an extremely complex history, such as the Californian margin. However, the task of the marine geologist is not so much to devise classifications which fit all possible instances, as to identify those processes which characterize various types of margins and set them apart from others.

2.7 The Shelf Areas

The submerged part of a continent is the *shelf.* We have seen earlier that emerged low-lying regions also can be argued to be part of the shelf. However, here we are concerned with the actual sea floor, covered by water.

Typically, shelves are flat and are not very deeply submerged. An average depth for the shelves rimming the Atlantic is near 130 m. Some shelves are quite wide, especially those on passive margins. These are, on the whole, depositional features, that is, they are built up by sediments. Narrow and rocky shelves, on the other hand, are common on active margins (see Fig. 2.6). Here erosional processes play an important role in shaping the shelf.

Some shelves extend deep into continents, and harbor *shelf seas* such as the Hudson Bay, the Baltic Sea, or the Persian Gulf. Most of the marine sediments found on land were originally deposited in shelf seas. To understand these sediments — which cover the greater portion of the continents — one needs to study sedimentary processes in modern shelf seas (see Chap. 3).

Quite generally, present shelf environments and sediment types show considerable variety over short distances. In part, this variability stems from the fact that the sea level stood much lower only 15,000 years ago. Conditions were entirely different then, and many portions of the present shelf still reflect those conditions in topography and sediment cover. The reason for the low sea level, of course, was the presence of continental ice sheets which locked up enough water to make the ocean go down by about 130 m. Sea level rise, due to the melting of the ice, was too rapid for the waves and currents to complete their work of smoothing the shelf during transgression, by erosion and deposition. Drowned river valleys, and their incompletely filled extensions across the shelves — valleys once cut by rivers running toward a more distant ocean — are reminders of the recent sea level rise. We shall return to this topic in Chap. 7.

We see that the nature of shelves reflects tectonics (active *versus* passive) and the recent rise of sea level, on a grand scale. On a regional scale, climatic conditions and sediment supply are of prime importance.

In the northern North Atlantic, for example, shelves everywhere show the effects of ice. The growth of ice not only led to exposure of the shelves, it also dumped enormous amounts of debris in places, the *moraines*. This material still sits on the shelves, off Newfoundland, and in the North Sea. The ice, moving far out onto the shelves, also actively carved deep ravines and depressions, which have not yet been filled. The fjords of Norway, Greenland, and western Canada are witnesses to this powerful action of the ice.

In tropical regions where the temperature never drops below 18° C, there are abundant shelves which are entirely built by carbonate-secreting organisms. The most varied shelf topography is that of coral reefs. Platforms alternate with deep channels, and irregular valleys are dotted with steeply rising mounds. Many a captain has lost his vessel in such waters. The great explorer of the Pacific, Captain James Cook (1728–1779), narrowly escaped this fate when trapped inside the Great Barrier Reef with his 368-ton Endeavour. His is a sailor's view of the irregularity of this type of shelf (Cook's Diary, 7-8-1770):

"After having well View'd our situation from the mast head I saw that we were surrounded on every side with Shoals and no such thing as a passage to Sea but through the winding channels between them, dangerous to the highest degree in so much that I was quite at a loss which way to steer when the weather would permit us to get under sail . . ."

Shelves formed by large deltas off river mouths (Amazon, Mississippi, etc.)

can be very flat and monotonous, in striking contrast to both the rugged ice-carved shelf and the irregular reef shelf. The rich supply of fine sediment associated with delta environments allows redistribution and smoothing by waves and currents. Of course, waves and currents can also build up dunes, barriers, beach berms, and sand waves, depending on circumstances (see Chap. 7). Examples for these conditions of high terrigenous sediment supply include the North Sea, the shelf off the Siberian rivers, the shelf of the Yellow Sea. A prime example is the Senegal delta where the shelf has less than 10 cm relief over several miles!

In any study of present shelves, we are chiefly faced with the question of how much of the observed morphology and sediment cover is inherited from past ice ages, and how much may be ascribed to the activity of the sea today. The task is complicated by the fact that "today" includes at least the last several hundred years. Within such time spans the sea can produce effects whose causes may not be obvious, especially if they include rare but powerful hurricanes and large earthquake-generated waves (tsunamis).

Tsunamis are produced chiefly in the trenches rimming the Pacific; they travel over thousands of miles in a few hours. They are very long waves, and so low in the open sea that they are not noticed on board a ship. When such a wave reaches a shelf area, however, it slows down and builds up, reaching enormous heights (tens of meters) where conditions are right. In this condition it can wreak havoc on the coasts of exposed shelves and also influence shelf morphology through the powerful bottom currents it produces.

2.8 The Shelf Break

The shelf break, where shelf joins upper slope, is a prominent morphological feature of continental margins. Its origin is somewhat of a mystery. In principle, the break marks the depth below which the influence of sea level on erosion and deposition wane rapidly (Chap. 7). The details are by no means understood, however.

The shelf break is commonly represented by a distinct increase in the slope at about 100 to 150 m. The global mean is near 130 m. In the Antarctic and Greenland it is very deep, down to 400 m. Here the break may it mark the depression by the ice load and the maximum depth of carving by ice. However, it is equally deep off southwest Africa, where this explanation does not work. Generally, the break in slope is distinct (Fig. 2.7), but it can also be gentle in places.

The fact that the shelf break is commonly between 100 to 150 m deep, strongly suggests that it marks the lowstand of glacial sea level. Apparently this lowstand was reached repeatedly in the maximum glaciations during the late Quaternary, thus exerting considerable control on shelf evolution. In any discussion of the shelf and the shelf break in a particular geographic area, the regional isostatic responses of the shelf to loading and unloading with water due to the changing sea level also must be considered. However, such responses are difficult to separate from regional uplift and subsidence.

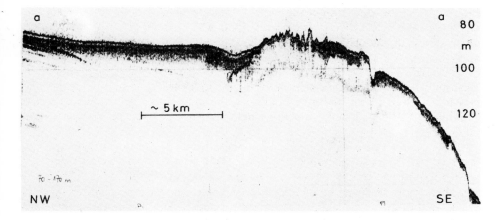

Fig. 2.7. Shelf break at the entrance of the Persian Gulf, subsurface echo-profile by the research vessel Meteor (1965). Note the accumulation of soft, layered sediment behind the rugged reef structure. Upper slope collects reefal debris. The reef is dead. Shelf break is somewhat above 100 m depth

2.9 Continental Slope and Continental Rise

The classic profile of Atlantic-type continental margins shows a steepening of the slope at the shelf break and a gradually diminishing descent toward the deep sea floor (Fig. 2.2). The relatively steep part below the shelf break is the *continental slope;* the gently descending part leading into the deep sea is the *continental rise.* The boundary between slope and rise is ill-defined. Perhaps we can say that the slope is definitely part of the margin, whereas the rise is built on oceanic crust and is essentially part of the deep sea environment.

Not all slopes and rises, of course, fit the textbook outline of the ideal Atlantic type — even in the Atlantic. Deep marginal ridges, as off Brazil, sheer walls of outcropping ancient sediments, and deep-lying plateaus such as the Blake Plateau off Florida can interrupt the ideal sequence.

The collision margins off Peru and Chile are characterized by a steep slope, without a rise — the trench takes care of the material which would normally build the rise. A descent in a series of steps is typical for the collision slope.

Rather complicated conditions prevail off much of the western U.S. coast. While the slope and rise off Northern California can be described simply enough in terms of coalescing deep sea fan deposits (Fig. 2.8 northern part), no such description is possible for the margin to the South. The Southern California Borderland looks much like an extension of the basin-and-range topography familiar from the Mojave Desert, into the sea. The tops of the ranges protrude as islands (Fig. 2.8b southern part).

The various physiographic diagrams commonly depicting the rapid descent from shelf edge to deep sea are somewhat misleading in that the slopes are in reality quite gentle: between 1° and 6° (see Table 2.1). A 1° slope, of course, would appear as a plain if one were standing on it.

47

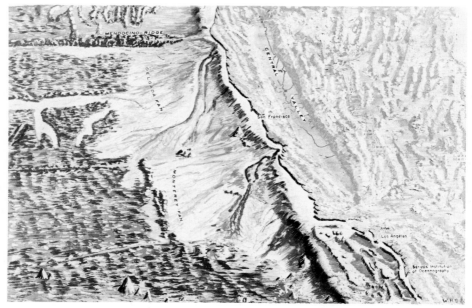

Fig. 2.8. Physiographic diagram of the ocean margin off California. Note the narrow shelves bounded by sea cliffs toward the land (uplift!). The Continental Borderland in the south is a submerged basin-and-range province. The continental slope essentially consists of enormous coalescing fans which transgress over the abyssal hills province. (Painting based on physiographic diagram of H. W. Menard, 1964)

From the variety of types of slopes it is clear that a number of different forces are at work in shaping them. We have referred to some of the endogenic forces when comparing Atlantic-type and Pacific-type margins. Changes in the thickness of continental crust, through remelting and assimilation into the mantle (subcrustal erosion) have been proposed as one mechanism influencing the evolution of margins. Such processes may be active, for example, in the Southern California Borderland.

Basically, most slopes are the surfaces of thick accumulations of sediment washed off the continents and mixed with marine materials of biologic derivation. The deep sea receives a rather miserly share of the continent's products: the bulk accumulates on the margins. The high rate of accumulation on many slopes leads to a precarious balance in places. When slopes get too steep and there is not enough time for solidifying the sediment, immense landslides can result. Earthquakes are likely to trigger such slides when conditions are right.

An example for a large-scale slide is shown in Fig. 2.9; the jumbled masses have come to rest at the foot of the slope and now form part of the continental rise. The seismic profile was obtained by echo-sounding *into the sea floor* with powerful "booms" of sound, rather than the "pings" used to define the surface of the floor. (The method is called *continuous seismic profiling.*)

The arrangement of the sedimentary layers below the sea floor in Fig. 2.9 shows that erosion has taken place in earlier times, alternating with periods of

48

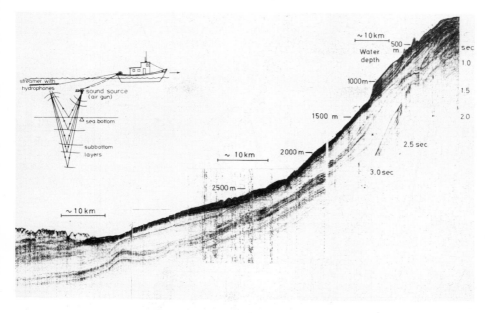

Fig. 2.9. Submarine mass movements off Dakar (NW Africa). Air gun record of Meteor Cruise 25/1971. Shelf edge *upper right.* Slide starts at 1050 m depth (return time for outgoing sound pulse: 1.4 s). Thickness of slide ∼200 m. Material came to rest below about 2300 m depth (=3.6 s) at the foot of the continental slope. *Insert left* air gun system with air gun as sound source. Acoustic signals are reflected by the sea floor and by subbottom layers and are recorded by hydrophones in the streamer

deposition. Erosion can be produced both by slides and by the action of strong deep currents flowing horizontally along the slope. Such currents have been dubbed *contour currents,* to distinguish them from the *turbidity currents* which are mud-laden bodies of water flowing downhill (Sect. 2.11 and 4.3.6). It is now thought that much of the sediment on the continental rises is carried there originally by turbidity currents and by slides starting somewhere near the shelf break, and that subsequently it is redistributed by contour currents.

2.10 Submarine Canyons

The continental slope is commonly cut by various types of incisions, ravines, and valleys, the most spectacular of which are the *submarine canyons.* The origin of these impressive features (Fig. 2.10) has long puzzled marine geologists and is still a matter of debate and intensive study.

Many large submarine canyons are very much like their counterparts on land: tributary systems in the upper parts, with meandering *thalwegs* (similar to river beds), and steep sides in places (20° to 25°, even 45°). Overhanging walls also occur (Fig. 2.11). As in river canyons, there is a continuous descent of the deepest

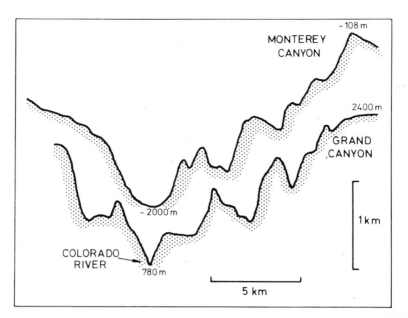

Fig. 2.10. Profile of Monterey Canyon compared with that of the Grand Canyon of Arizona. (From F. P. Shepard and R. F. Dill, 1966). The resemblance is coincidental, but illustrates the enormous size of the Monterey Canyon (see Fig. 2.8)

Fig. 2.11. Illustration of morphological similarities of subaerial canyon (Grand Canyon, *left*) and submarine canyon (La Jolla Canyon, *right*). Note steepness and overhang in both cases. (Photo left, E. S.; underwater photo courtesy R. F. Dill)

point of the valley, with maximum slopes of 15% near the coast and gentle slopes of about 1% farther out to sea.

Quite commonly, canyons cut right through the shelf and may line up with valleys on land. Ancient fishing villages are sometimes located at the heads of submarine canyons: the greater water depth disperses of high waves away from the coast and provides some safety for boats and houses on shore. Examples are Nazare in Portugal, and Kayar in Senegal. Some canyons, however, only start at the edge of the shelf. The seaward extension of submarine canyons can reach to the end of the continetal rise and beyond, and run for hundreds, even thousands, of kilometers across the deep sea floor (Fig. 2.2b, Hudson Canyon; Fig. 1.3, Mid-Ocean Canyon, off Eastern Canada).

Canyons and various other types of slope valleys can be extremely abundant in places, such that the slope is dissected like the remnants of a mesa on land. Nor do those canyons necessarily run straight downhill; they may be cut at an angle to the slope. In other places, submarine canyons seem to be absent, presumably by reason of low sediment supply (insufficient for producing downhill currents) or because of gentle slopes, or both. The best-known canyons are associated with the mouths of large rivers — the Congo, Indus, Ganges and the Hudson. In essence, they are valleys within the cones of debris brought by the rivers, maintained by erosion in the upper reaches, but acting as distributaries in the lower parts. Here they are bordered by levees.

There are a large number of hypotheses which were put forward over the years to account for the origin of submarine canyons. In fact, different types of canyons must have different origins. For example, it has been demonstrated through deep sea drilling, that the Mediterranean became isolated from the world ocean some 5 to 6 million years ago and dried up during periods of evaporation. At those times, deep canyons could have been carved by the familiar action of rain and run-off, into the margins of the Mediterranean. Indeed, the true floor of the Nile Valley is very deep, supporting the idea of canyon cutting during dessication. The present river runs on top of a thick pile of sediment, which has filled the canyon since.

In principle, every ocean basin with salt deposits in its margins presumably had a low sea level at one time, allowing subaerial cutting of canyons. This includes the North and the South Atlantic. However, we can hardly invoke such drastic falls of sea level everywhere in the world ocean. Thus, there must be a way to make submarine canyons by cutting them under water. The fact that so many canyons are off river valleys suggests a mechanism: some sort of submarine river flowing on the sea floor. This submarine flow cannot be an extension of the river entering the sea: its water is fresh and therefore less dense than seawater. River water floats on seawater. For example, surface ocean salinities are low for a hundred miles out to sea off the mouth of the Amazon, mightiest of all rivers. But seawater with a high mud content is heavy enough to flow downhill on the ocean floor.

One way to stir mud into the water is to have slides start in the soft sediments deposited off river mouths. Large amounts of mud are brought down the river during floods, and such mud is rather unstable. Hurricanes and wave action may produce large amounts of heavy, muddy water through stirring up of the sediment. Earthquakes may be agents for starting mudslides which turn into muddy downhill flow, or *turbidity currents.*

Such currents, deriving their energy from gravity, could conceivably erode submarine canyons in the steep part of the slope. If so, both the material originally

51

suspended and the material eroded, must be deposited in the areas where the slope is insufficient to maintain the power of the currents, as the gravity force wanes. We should find these deposits, and we do.

2.11 Submarine Fans

Steep-walled canyons, such as illustrated in Fig. 2.11, typically cut the upper continental slope. Further downslope their character changes: they grade into fan valleys which traverse enormous semi-conical sediment bodies, the *deep sea fans* (Fig. 2.12). Such fans are exactly what we expect to see if the submarine canyons are funnels for turbidity currents (see also Sect. 4.3.6 and 8.5).

Turbidity currents, then, are sea-floor-hugging, mud-laden bodies of water driven down-hill by gravity. They are envisaged as the product of brief catastrophic events, triggered by floods, storm action or earthquakes. Normally,

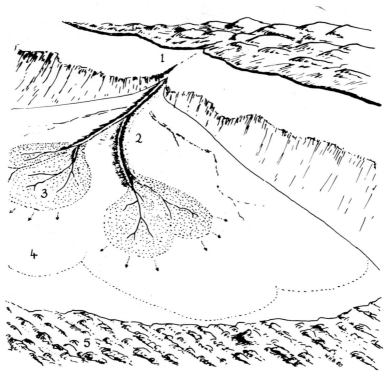

Fig. 2.12. Build-up of deep sea fans. *1* Canyon cutting through shelf and upper slope, traps and funnels sediment to the fan; *2* Upper fan valley, with levees preventing escape of turbidity current. Levees can be breached (note dead channels); *3* Active suprafan with distributary channels; *4* Outer fan, receiving fine spill-over sediment from suprafan; *5* Abyssal hill region beyond fan. Valleys between hills may have distal fan material. (Based on a sketch by W. R. Normark, 1970, Am Assoc Pet Geol Bull 54: 2170)

nothing much is happening in submarine canyons except for the back and forth of tidal currents, as was shown by measurements on currents done by F. P. Shepard and collaborators. But from time to time, perhaps once every thousand years, a great muddy flood tears down a canyon, and dumps its load somewhere on the fan. The occasional floods in desert regions, which build fans along the valley sides, are analogous phenomena.

The deposits of turbidity currents are called *turbidites;* we find them both within slope sediments and in the *abyssal plains* (Sect. 8.5). Depending on their size and duration, turbidity currents rushing down a fan valley leave the distributary channels at various points, and build up turbidites as their velocity slows (Fig. 2.12).

Most turbidites are thin and are soon destroyed through reworking by bottom-living organismus and by bottom currents. Thick layers, of course, can survive this process and are then recognizable in the sediment sequence.

It reflects the youth of the science of marine geology that this major process of *sediment transport and erosion,* which is responsible for much of ocean margin morphology, was essentially unknown until the 1950's. It took yet another decade for the *tectonics* of ocean margins to come into focus, through the theory of sea floor spreading. The history of ocean margin development is still, on the whole, poorly understood because of the lack of samples from within the sediment wedges: recovery by deep drilling is now possible, but is extremely expensive.

3 Sources and Composition of Marine Sediments

3.1 The Sediment Cycle

Marine sediments show great variety: there is the debris from the wearing down of continents, the shells and organic matter derived from organisms, the salts precipitated from seawater, the volcanic products such as ash and pumice (Fig. 3.1).

Most sediments ultimately derive from the *weathering* of rocks on land. The action of ice and water, of hot and cold, breaks up the rocks into small particles and leaches them, destroying the more soluble minerals. A special type of "weathering" occurs at the ridge crests and other young volcanic features in the sea: the reactions of heated seawater with basalt. Such reactions may contribute considerable amounts of matter to seawater. Just how much is not known, however.

The sea itself takes a toll from the continents: waves and tides can eat into land, taking debris offshore. The material may stay on the shelf, or it may bypass the shelf to accumulate on the slope or below.

Sediment which stays on the shelf, of course, essentially stays on the continent. The other sediment, eventually, is carried back to the continents or to the mantle by sea floor spreading, completing the cycle.

3.2 Sources of Sediment

3.2.1 River Input. The dissolved and particular load of rivers constitute the main source of sediment: the coalescing fans which build the continental slope off California (see Fig. 2.8), for example, are largely riverine mud, with an admixture of shell and organic matter. These precipitates (calcium carbonate, $CaCO_3$; and opal, $SiO_2 \cdot nH_2O$) also derive largely from river input, namely from the dissolved load of rivers entering the ocean. We can make a rough estimate of how much material enters the sea. Deep sea sedimentation rates are between 1 and 20 mm per 1000 years. Slope sediments accumulate at a rate of up to 100 mm per 1000 years, approximately.

Let us take the value of 100 mm/1000 years for 10% of the ocean, and a value of 5 mm/1000 years for the rest: the average is near 15 mm/1000 years $(0.1 \times 100 + 0.9 \times 5)$. The ocean floor covers more than twice the area of the land. Hence, the continents — if they are the ultimate source of all sediment — must wear down at a rate of near 30 mm/1000 years. Not a bad guess, actually. Recent

54

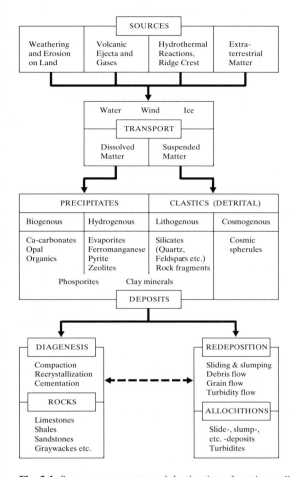

Fig. 3.1. Sources, transport, and destination of marine sediments

estimates of erosion of the U.S. portion of North America suggest 65 mm/1000 years. Of course, man's agricultural activities have greatly increased erosion rates, although we may not yet be able to define this effect unequivocally.

Another rough calculation may be based on an estimated sediment supply from rivers of 12 km^3/year. If we distribute this amount on the 362 million km^2 of sea floor, we get a sedimentation rate of about 30 mm/1000 years, and a corresponding erosion rate of 60 mm/1000 years for the continents.

In general, mechanical weathering dominates in high latitudes, where water acts mostly as ice, and in deserts, where water is subdued as an agent. Chemical weathering (leaching) is favored by high rainfall and high temperatures and

55

dominates in tropical areas. In extrapolating these and other present-day patterns into the past we must always remember that we live in a highly unusual period. High mountain ranges and the powerful abrasive action of the continental ice masses have greatly increased mechanical erosion for the last several million years.

3.2.2 Glacier Input. The great importance of ice in delivering sediment to the sea in high latitudes is readily appreciated when contemplating the immense masses of outwash material they bring to the shores, to be reworked on the shelves. Less important as concerns sediment mass, but very interesting for paleoclimate reconstruction, is the fact that calving glaciers can transport both fine and very coarse material far out to sea. When the icebergs melt they drop their load. Around Antarctica this type of transport reaches to about 40° S. In the North Atlantic the transport boundary roughly follows the present boundary between very cold and temperate waters (Fig. 3.2). During the last ice age this limit extended much further south, to a line between New York and Portugal (Sect. 7.2.3). At present about 20% of the sea floor receives at least some ice-transported sediments.

3.2.3 Input from Wind. In contrast to ice transport, wind can only move the fine material. Medieval Arabian scientists had noted the dust coming out of the Sahara into the "dark sea" of the Atlantic Ocean (Fig. 3.3). In the 1800's, Charles Darwin assumed (correctly) that the dust must build up the ocean floor. As the dust blows out to sea, the larger particles settle out first and the grain size becomes continuously finer. Particles from a large Saharan dust storm, in 1901, had an average size of about 0.012 mm in Palermo, and 0.006 mm in Hamburg. (This would be classified as extremely fine silt.) During this same storm, up to 11 g of dust per m^3 were measured over the Mediterranean.

   The rate of dust fall from the air can best be measured in snowfields and ice cores with annual layering. Even in the Antarctic and in Greenland — far from desert sources — the rate is quite appreciable: 0.1 to 1 mm dust in 1000 years. Exactly how much dust is falling on the sea floor is not known. Some estimates suggest that much or most of the deep sea clay is derived from wind input. Such clay accumulates at between 1 mm/1000 years in the North Pacific, and 2.5 mm/1000 years in the Atlantic.

3.2.4 Volcanic Input. A substantial amount of material is delivered by volcanoes, especially those associated with active oceanic margins. In fact, the composition of deep sea clay (Chap. 4) suggests that in the time before 10 million years ago, before mountain building and glaciation drastically changed conditions, the main source of deep sea clay in the Pacific was the decomposition of volcanic ash. Of course, volcanic material also is eroded from land and brought into the sea as

56

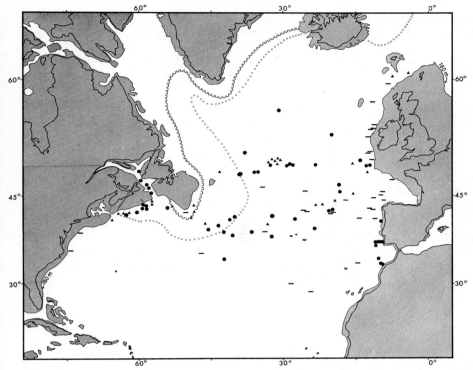

Fig. 3.2. Distribution of ice-rafted materials in the central North Atlantic. Present limits of drift ice (normal and extreme) run from around Newfoundland toward Greenland and Iceland. During the last ice age, this limit went from New York straight across to Portugal. *Triangles* surface samples; *dashes* dredge samples; *circles* core samples. (After H. R. Kudrass, 1973, Meteor Forschungsergeb Reihe C 13: 1)

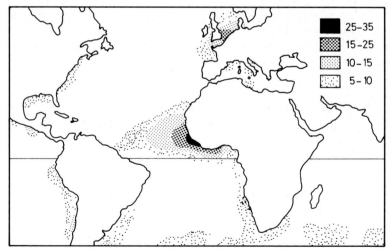

Fig. 3.3. Abundance of haze from dust, over the Atlantic Ocean. (From G. O. S. Arrhenius, 1963, in M. N. Hill, Sea 3: 695)

57

terrigenous sediment. Subtle differences in chemistry, "chemical fingerprints", hold the clue as to which is the source for a given volcanogenic deposit.

Geologically speaking, volcanoes have a short life, and single eruptions are flash-like events. Therefore ash layers can be used for regional stratigraphic purposes *(tephrachronology)*, for example, in the Mediterranean or around Iceland (Fig. 3.4). Volcanic activity also adds gases and hydrothermal solutions to the ocean. These fluxes have an important bearing on the evolution of the chemistry of ocean and atmosphere and are the subject of intense study at present.

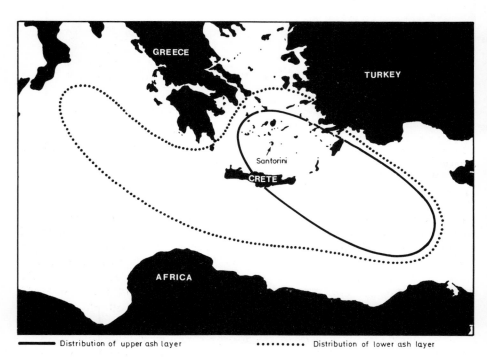

━━━━ Distribution of upper ash layer   •••••••••• Distribution of lower ash layer

Fig. 3.4. Distribution of volcanic ash produced by two large volcanic explosions in the Aegean Sea, presumably of Santorini. The lower ash layer marks a prehistoric event (>25,000 yrs). The upper layer is less than 5000 yrs old; the volcanic explosion creating it may be the one which brought catastrophe to the Minoan culture, 3450 years ago. (After D. Ninkovich, B. C. Heezen, Nature [London] 213: 1967, 582, in K. K. Turekian, 1968, Oceans. Prentice-Hall, New Jersey)

3.3 Sediments and Sea Water Composition

3.3.1 Acid-Base Titration. To a first approximation, sea water is a solution of sodium chlorine (NaCl), that is, table salt. Sodium and chloride make up 86% of the ions present, by weight (Table 3.1). The other major ions are magnesium, calcium, and potassium (alkaline and earth alkaline metals), and the acid radicals

58

Table 3.1 Comparison between sea water and river water

Ion	Sea water			Average river water		
	ppm[b]	Weight %	Rank	ppm	Weight %	Rank
Cl^-	18,980	55.0	(1)	7.8	6.4	(5)
Na^+	10,561	30.6	(2)	6.3	5.2	(6)
SO_4^{2+}	2,649	7.7	(3)	11.2	9.3	(4)
Mg^{2+}	1,272	3.7	(4)	4.1	3.4	(7)
Ca^{2+}	400	1.1	(5)	15.0	12.4	(2)
K^+	380	0.4	(6)	2.3	1.9	(8)
HCO_3^-, CO_3^{2-}	140	0.2	(7)	58.8	48.6	(1)
Br^-	65	0.1	(8)	0.02	–	
H_3BO_3	26	–		0.1–0.01	–	
Sr^{2+}	13	–		0.09	–	
F^-	1.4	–		0.09	–	
H_4SiO_4	1	–		13.1	10.8	(3)
NO_3^-	0.5	–		1.0	0.8	
Fe^{2+}, Fe^{3+}	0.01	–		0.67	0.5	
$Al(OH)_4^-$	0.01	–		0.24	0.2	
Sum	34,479	=100%		120.8	=100%	

[a] From D. A. Livingstone DA (1963) U.S. Geol Surv Profess Paper, 44 DG: ("Hard" water contains roughly twice the average total, "soft" water one half)
[b] ppm=parts per million (g per ton)

sulfate and bicarbonate. Since the major cations form strong bases, but bicarbonate forms a weak acid, the ocean is slightly alkaline, with a pH of near 8, rather than the point of neutrality (pH=7). (pH is a measure of the acidity of a solution.) On the whole, the salty ocean may be understood as the product of emission of acid gases from volcanoes (hydrochloric, sulfuric, and carbonic acid) and the leaching of common silicate rocks whose minerals have the form [Me Si_a Al_b O_c] where Me stands for the metals Na, K, Mg, and Ca, and the remainder makes insoluble silica-aluminum oxides, that is, clay minerals.

How stable was the composition of seawater through geologic time? If we use the above concept of acid-base titration, and assume that seawater is a solution in equilibrium with the sediments on the sea floor, the result is that the composition should be rather stable. Also, we can obtain samples of seawater salt in the ancient salt deposits. Their composition indicates that sea salt did not vary much over the last 600 million years: probably less than a factor of 2 for any one major component. Paleontologic evidence certainly agrees with this assessment. Already in the early Paleozoic there are organisms whose closest modern relatives have rather narrow salt tolerances: radiolarians, corals, brachiopods, cephalopods, echinoderms. Of course, adaptation of the organisms to an increasing or changing salt content cannot be ruled out.

On comparing the average composition of river water with that of seawater, one notes drastic differences. Essentially, river water is a solution of calcium bicarbonate and silicic acid, with a small admixture of the familiar salts — a substantial proportion of which are recycled marine salts. We see from this discrepancy in the composition of seawater and river water that the river influx per se is irrelevant to the make-up of sea salt. What counts is the *solubility* of the salts. In the simplest terms, the soluble salts are abundant in seawater and the others are not.

3.3.2 Residence Time. For steady-state conditions, output must equal input. Thus the seawater has to rid itself of all new salt coming in, in the same proportions as they are added. Where are the "sinks" for this material? The quantitative assessment of sinks is a major geochemical problem. For calcium carbonate the sink is calcareous skeletons built by organisms; for silica it is opaline skeletons. The metals presumably leave the ocean in newly formed minerals such as authigenic clay, oxides, and sulfides, and in zeolites, as well as in alteration products resulting from reactions between hot basalt and sea water, at the ridge crest. The sulfur is precipitated in heavy metal sulfides in anaerobic sediments near the land. Some salt leaves with the pore waters in the sediments. Under the assumption that the ocean does not change its composition, we can calculate the average time a seawater component remains in the water, before going out as sediment. This time is called *residence time.*

Calculating the residence time is analogous to figuring out how long people will stay in an exhibit hall: count the people present, and the number entering per unit time. The ratio is the average viewing time. Similarly,

$$t = A/r \tag{3.1}$$

where A is the amount present, and r is the input. Some residence times are given in Table 3.2. Clearly. so, sodium and chloride have a long residence time, and silica has a very short one. The residence time is, in essence, a measure of the geochemical solubility (or inversely, reactivity) of the substance in question. Equation (3.1) was once used to calculate a "salt age" for the ocean, under the assumption that the ocean started out fresh and retained all sodium since. This salt age came out near 100 million years. In our earlier analogy, calculating the salt age corresponds to figuring the time the exhibit opened, from the number of people present and the rate at which they are coming in. The salt age was once useful as a minimum estimate for the scale of geologic time.

Table 3.2 Residence times[a] of some elements in sea water (in millions of years)

Na	210
Cl	>200
Mg	22
K	10
Ca	1
Si	0.04

[a] The residence time ist the ratio between the mass within the reservoir and the mass introduced per year. (Data mainly from E. D. Goldberg ED 1965, in Chemical Oceanography vol. 1, 163–196, Academic Press, New York)

3.4 Major Sediment Types

There are essentially three types of sediments: those that come into the ocean as particles are dispersed, and settle onto the sea floor; those that are precipitated out of solution directly; and those that are made by organisms. For convenience we may call the first type *lithogenous,* the second *hydrogenous* and the third, *biogenous* (Fig. 3.5); Box 3.1). Around the ocean margins, lithogenous sediments are predominant, although there also are salt deposits, and biogenic sediments. In the deep sea, biogenous sediments predominate, especially *calcareous ooze.*

a

b

Fig. 3.5a, b. Familiar examples of two major types of beach sand: **a** lithogenous (La Jolla) and **b** biogenous (Hawaii). (Photos W. H. B.)

Box 3.1 Classification of marine sediment types

Lithogenous Sediments. Detrital products of disintegration of pre-existing rocks (igneous, metamorphic, or sedimentary, see Appendix A6) and of volcanic ejecta (ash, pumice). Transport by rivers, glaciers, winds. Redistribution through waves and currents. Nomenclature based on *grain size* (gravel, sand, silt, clay). Additional qualifiers are derived from the *lithologic components* (terrigenous, bioclastic, calcareous, volcanogenic etc.), and from the *structure* and *color* of the deposits.

   Typical examples are as follows (environment in parentheses):
   Organic-rich *clayey silt,* with root fragments (marsh)
   Finely laminated *sandy silt,* with small shells (delta-top)
   Laminated quartzose *sand,* well sorted (beach)
   Olive-green homogeneous *mud* rich in diatom debris (upper cont. slope) (*mud* is the same as terrigenous clayey silt or silty clay)
   Fine-grained lithogenous sediments (which become *shale* upon aging and hardening) are the most abundant by volume of all marine sediments (about 70%). This is largely due to the great thickness of continental slope sediments.

Biogenous Sediments. Remains of organisms, mainly carbonate (calcite, aragonite), opal (hydrated silica), and calcium phosphate (teeth, bones, crustacean carapaces) (see Table 3.3). *Organic sediments,* while strictly speaking biogenous, are commonly treated separately. Arrival at the site of deposition by in situ precipitation (benthic organisms living there), or through settling via water column (pelagic organisms; coarse shells fall singly, small ones commonly arrive as aggregates). Redeposition by waves or currents. Redissolution common, either on sea floor or within sediment. Nomenclature is based on *type of organism* and also on *chemical composition.* Additional qualifiers from *structure, color, size, accessory matter.* Typical examples:
   Oyster bank (lagoon or embayment)
   Shell sand (tropical beach)
   Coral reef breccia (slope below coral reef) (*breccia* is coarse broken-up material, in this case reefal debris)
   Oolite sand, well sorted (strand zone, Bahamas)
   Light grey calcareous ooze, bioturbated (deep-sea floor)
   Greenish grey siliceous ooze (deep-sea floor)
   Biogenous sediments are widespread on the sea floor, covering about one half of the shelves and more than one half of the deep ocean bottom, for a total of 55%. About 30% of the volume of marine sediments being deposited at the present time may be labeled biogenous, although they may have considerable lithogenous admixture.

Hydrogenous Sediments. Precipitates from seawater or from interstitial water. Also products of alteration during early chemical reactions within freshly deposited sediment. Redissolution common. Nomenclature based on *origin* ("evaporites") and on *chemical composition.* Additional qualifiers from *structure, color, accessories.* Typical examples:
   Laminated translucent halite (salt flat)
   Finely bedded anhydrite (Mediterranean basin, subsurface)
   Nodular greyish white anhydrite (ditto)
   Manganese nodule, black, mammilated, 5 cm diam. (deep Pacific)
   Phosphatic concretion, irregular slab, 15 cm diam., 5 cm thick, light brown to greenish, granular (upwelling area)
   Hydrogenous sediments, while widespread (ferromanganese!) are not important by volume at present. At times in the past, when thick salt deposits were laid down in the newly opening Atlantic, in the Mesozoic, and much later in a dried-up Mediterranean (end of Miocene) the volume of hydrogenous sediments produced was considerable. The salinity of the ocean may have been appreciably lowered during those times.

3.5 Lithogenous Sediments

3.5.1 Grain Size. The bulk of sediment around the continents consists of debris washed of the continents. This debris resulted from the mechanical break-up, with or without leaching, of continental igneous and sedimentary rocks. The product is fragments of rocks and minerals. One property that is extremely important for assessing source and transport processes is *grain size* (Sect. 4.1). One distinguishes *gravel, sand, silt,* and *clay*. Thus *sand,* for example, is all material of a size between 0.063 mm and 2 mm, regardless of its composition or origin. Likewise, *clay* is everything below a particle size of 0.004 mm (or 4μm). This terminology is sometimes confusing, because "clay minerals" are minerals of a certain type, abundant in clay-type deposits, but they are not invariably of "clay size".

Overall, there is a gradation of grain sizes from source to the place of deposition, such that the bigger, heavier particles remain closer to the source while the clay-size particles can be carried over large distances. *Gravel* then, does not commonly travel far, except when taken along by ice (which does not care much about the size of its load). Since gravel consists of pieces of rock which can be identified and matched with the source area, it is commonly possible to determine the origin of a gravel assemblage. Except on sea floor supplied by glaciers or by coral reef, gravel is not an important constituent of marine sediments.

3.5.2 Sand is typical of beach and shelf deposits (Fig. 3.5). Like gravel, it may consist of rock fragments, a striking example being provided by black beach sands on Hawaii, which are made of volcanic rock. Commonly, however, sand consists of fragments of the minerals quartz, feldspar, mica, and others. Except for quartz, these are compounds of the type [Na, K, Mg, Ca]$_a$ [Si, Al]$_b$ O$_c$, that is, alumino silicates. Quartz (SiO$_2$) is the mineral most resistant to abrasion and to leaching, and sands which have been extensively reworked are therefore greatly enriched in quartz (mature sands). The first minerals to be destroyed during weathering, transport, and reworking are the iron-rich minerals. Feldspar also is not very resistant to chemical destruction.

In many tropical beaches, sand may consist entirely of fragments of calcareous skeletons, of molluscs, corals, and algae. One might argue whether a deposit made up of such "bioclastic" material should be counted as "lithogenous", as derived from pre-existing rocks or as "biogenous". Strictly speaking, there is a cycle of mechanical break-up and abrasion involved (hence lithogenous), but for geochemical balance calculations we might include this material with chemical deposits. Carbonate particles are easily abraded and chemically eroded. Thus, in a mixture of carbonate and quartz sand, the quartz will soon dominate when the sand is reworked.

The source areas and dispersal history of sands can also be explored by noting the compositional types of "heavy minerals" (densities greater than 2.8 g/cm^3; examples: the silicates hornblende, pyroxene, olivene, and also magnetite,

ilmenite, rutile). The *heavy mineral association* allows the mapping of depositional provinces, which in turn provides clues to the action of shelf currents and
other factors (Fig. 3.6).

The *shape* of sand grains, especially quartz grains, can give clues to their
origin. For example, grains in glacial deposits have sharp edges, whereas
reworking by waves leads to rounding. Dunes especially collect well-rounded
grains, and such grains may be "frosted". One problem with a straightforward
application of these concepts is that quartz grains are so resistant, they can be
recycled many times by erosion and redeposition of old sedimentary deposits
(polycyclic sands). Another problem is that etching of grains can take place *after*
deposition, within the sediment, obliterating and confusing the surface markings.

3.5.3 Silt-sized sediment is very characteristic of continental slope and rise, but
may occur at any place on the shelf where conditions are quiet, so that it is not
washed out by wave or current action. The composition of the silt (0.063 mm to
0.0004 mm) is much like that of sand in the coarser end of the range, and like that
of clay in the finer end. Mica is especially abundant in terrigenous silt.

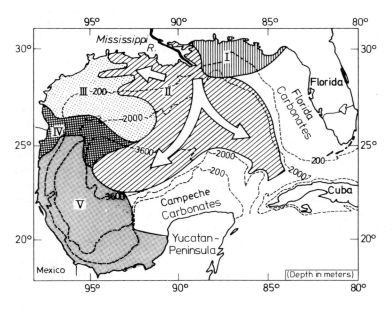

Fig. 3.6. Heavy mineral provinces of the Gulf of Mexico, based on typical mineral associations. *I* East
Gulf; *II* Mississippi; *III* Central Texas; *IV* Rio Grande; *V* Mexico. Carbonate particles are dominant
off Mexico and Yucatan. The heavy mineral patterns contain clues about the sources and transport
paths of terrigenous sediments. (After D. K. Davies, W. R. Moore, 1970, J Sediment Petrol 40: 339)

Sand is commonly studied with a binocular microscope, and the composition of clay is investigated by X-ray diffraction. The study of silt traditionally fell into the crack between these two methods and has had rather a low popularity rating. More recently, the scanning electron microscope has made it possible to investigate this size fraction in more detail (Fig. 3.7). The composition of the silt is usually closely related to that of the associated fine sand fractions.

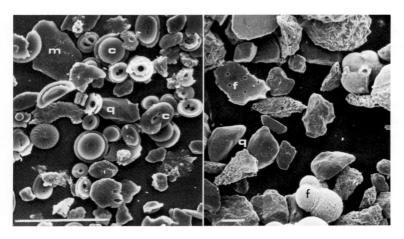

Fig. 3.7. Recent hemipelagic sediment from Continental Rise off NW Africa (Cape Verde Rise). SEM photos, scale bar = 20 μm. *Left* fine-silt fraction (2–6 μm) predominantly composed of coccoliths *(c)* and of some detrital mica *(m)* and quartz *(q)*. *Right* coarse silt fraction, mainly detrital grains *(q* quartz) and foraminifera tests and fragments *(f)*. (Photo courtesy D. Fütterer, Kiel)

3.5.4 Clay-sized sediment is ubiquitous on shelves, slopes, and in the deep sea. Like the silt, it indicates *low energy environments* where abundant, because it is easily picked up and washed away. Exceptions are clays in those high energy environments where clay supply is extremely high, as off some tropical rivers. Much of the clay, as mentioned, consists of clay minerals, products of weathering that are brought in by rivers and winds and are dispersed by waves and currents.

The common clay minerals, or clay mineral groups, are montmorillonite, illite, chlorite, and kaolinite. Their distribution and paleoclimatic significance will be discussed in more detail in Chap. 8.

The high surface areas of clay particles give clayey sediments special chemical properties. For example, they readily adsorb a large variety of substances, and react with the ions in seawater and in interstitial waters. Thus, during *diagenesis,* within buried sediments, new clay minerals can form slowly. Ultimately this has important implications for the chemistry of seawater and for geochemistry in general.

Clayey sediments in regions of high sedimentation rate are commonly rich in organic matter, partly because organics cling to the clay during deposition, and

65

partly because where conditions are quiet enough for the desposition of clay they also are favorable for the deposition of the fluffy organic particles. Much of the marine clays may actually be brought to the floor within fecal pellets of organisms filtering the water. Such clay is associated with food particles. Recent experiments with sediment trapping equipment off California and elsewhere suggest that this "fecal pellet transport" mechanism is a significant agent of sedimentation. The association between clay and organic matter is important in the search for source rocks, in petroleum exploration.

3.6 Biogenous Sediments

3.6.1 Types of Components. Organisms produce sediments in the form of skeletal materials and organic matter. The label "biogenous" generally refers only to calcareous, siliceous, and phosphatic hard parts.

Virtually all of the calcium carbonate deposited on the sea floor belongs in the category of biogenous sediments. On the shelf it is chiefly the shells and skeletons of benthic organisms, while on the slope there is an increasing proportion of planktonic remains away from the coast. In the deep sea, finally, plankton remains are entirely dominant. Opaline skeletons are also important — again mainly benthic remains (sponges) on certain shelves, and planktonic ones (diatoms, radiolarians) in slope sediments and in the deep sea (see Table 3.3). In addition to carbonate and silica, phosphatic particles are produced by organisms. The sedimentation of such particles plays an important role in the phosphorus budget of the biosphere, and is geochemically of great interest. The various other types of hard parts — strontium-sulfate, manganese-, iron-, and aluminum-compounds — are intriguing from an evolutionary standpoint but do not noticeably contribute to sediment patterns. Organic matter (organogenic sediment) will be taken up in Chapter 10, when discussing hydrocarbons.

3.6.2 Benthic Organisms are responsible for building up enormous masses of shelf material off certain coasts. We have earlier referred to the Great Barrier Reef, which is composed of calcareous shells and skeletons of corals, algae, molluscs, and foraminifera. Much of Florida sits on top of shelf carbonate built by benthos. Both aragonite and calcite is secreted by various types of organisms (see Table 3.3). The content of magnesium varies between groups of organisms, but also within the same group. There is a tendency for higher magnesium values in those skeletons precipitated in warmer water.

The great importance of carbonate-secreting benthos in producing sediment is apparent all through the Phanerozoic record, that is, the last 600 million years or so. In the Tertiary, we have immense masses of shelf carbonates building montain chains in the Alps and Himalayas. Platform carbonates were deposited all around the *Tethys,* the seaway which once linked the western Pacific and the Atlantic through Asia and Europe. In the Mesozoic, we have extensive deposits of shelf carbonate also: the oil fields in Arabia and in Mexico are associated with

Table 3.3. Inorganic constituents and carbonate minerals in hard parts of marine organisms (for systematics see Appendix A9)

Bacteria	Aragonite ($CaCO_3$)
Diatoms	Opal ($SiO_2 \cdot nH_2O$)
Coccolithophores	Calcite ($CaCO_3$)
Chlorophyta	Aragonite
Rhodophyta	Aragonite, Mg-calcite
Phaeophyta	Aragonite
Foraminifera	Calcite, Mg-calcite, aragonite (rare)
Radiolarians	Opal, celestite ($SrSO_4$)
Sponges	Mg-calcite, aragonite, opal, celestite (rare)
Corals	Aragonite, Mg-calcite
Bryozoans	Aragonite, Mg-calcite + aragonite
Brachiopods	Calcium-carbonate-phosphate, calcite
Echinoderms	Mg-calcite
Molluscs:	
Gastropods	Aragonite, aragonite + calcite
Pelecypods	Aragonite, aragonite + calcite, calcite
Cephalopods	Aragonite
(Various phosphate minerals and iron-oxides also have been detected in certain mollusks)	
Annelid worms	Aragonite, arag. + Mg-calcite, Mg-calcite
Arthropods:	
Decapods	Mg-calcite, amorphous Ca-phosphate
Ostracods	Calcite, aragonite (rare)
Barnacles	Calcite, aragonite (rare)
Vertebrates	Ca-phosphates

cavernous Mesozoic reef limestones, again built up within the Tethys seaway. In the Paleozoic, shelf carbonates built the Permian reefs in the Rockies, which contain a rich fauna of ancient corals, brachiopods, and molluscs. The carbonates of the Caledonian mountains running from Norway to Scotland and Wales into Newfoundland and Appalachia mark an ancient Atlantic seaway (Fig. 1.20).

The shallow water limestones of the geologic record, produced in the shelf environments of ancient oceans, commonly contain an admixture of siliceous rocks, as layers or nodular masses of *chert* arranged along horizons parallel to the bedding.

Mineralogically, this material, in the main, is finely crystalline quartz; in the Tertiary it may be *cryptocrystalline*, or it may be still amorphous in parts. The origin of these flinty masses has long been a mystery (although this did not prevent their use for tool-making in the Stone Age, of course). In the Miocene Monterey Formation of California, the source of the silica is in the shells of planktonic diatoms. However, in shelf limestones, the source may well have been siliceous sponges. Such sponges are abundant on shelves and upper slopes in the present ocean wherever there are silica-rich waters, that is, in high latitudes and in upwelling areas.

Why is there no evidence for incipient chert formation in modern shelf carbonate areas?

The reason is that silica concentrations are very low in present tropical waters. Diatoms and sponges do precipitate silica, but their skeletons are delicate and are quickly redissolved in the highly undersaturated seawater, which is stripped of its silica by diatoms in upwelling regions. Could ancient oceans have had higher concentrations of silica in the water? This is quite possible, if such oceans were much less fertile than the present ones. Also, it has been suggested that a high level of volcanic activity favored the formation of siliceous sedimentary rocks by supplying large amounts of silica. For the Eocene this idea has been offered to explain the abundance of cherts in deep sea carbonates. These cherts caused considerable difficulties for deep sea drilling, since they quickly destroy the drill-bit, and impede recovery of sediment cores.

3.6.3 Planktonic Organisms contribute a significant amount of material to slope sediments (Fig. 3.7). Deep-sea carbonates are made almost entirely of planktonic remains (see Chap. 8). Shelf sediments, especially those of the geologic past, also can contain considerable amounts of planktonic remains. For example, the English chalk is made largely of the remains of planktonic calcareous algae, the *coccolithophorids* (which produce the little platelets called *coccoliths*). The role of diatoms in supplying opal to margin sediments has been mentioned.

In general, the remains of planktonic organisms increase relative to those of benthic organisms in the offshore direction. A well-known application of this principle is the determination of the ratio of planktonic to benthic foraminifera in the mapping of marine facies. In the deep sea the ratio is greater than 10 to 1. On the edge of the shelf it is close to 50:50. In the Persian Gulf, a somewhat restricted shelf sea, the plankton-benthos ratio is even lower: about 3 to 7 near the entrance, and less than 1 to 10 in the interior. Thus, the planktonic remains are typical for the open ocean.

3.7 Non-skeletal Carbonates

3.7.1 Carbonate Saturation. In the present ocean carbonate precipitation is either within organisms (shells, skeletons) or is associated with their metabolic activity (algal crusts). This need not always have been so: a certain proportion of the limestones and associated carbonates in the geologic record may have been precipitated inorganically.

Where would one look for such inorganic precipitation today? To guide us in the search for likely places, we need to consider briefly the chemistry of carbonate precipitation and dissolution.

Seawater which spontaneously precipitates a mineral — for example aragonite, $CaCO_3$ — is said to be *supersaturated* with this mineral phase. Seawater which dissolves the mineral is *undersaturated*. When precipitation just equals dissolution, *saturation* obtains, that is, the solution is *in equilibrium* with the solid. The degree of saturation is expressed as the ratio of the ionic product of reactants present to the product necessary for saturation (brackets indicate concentrations):

$$D_{sat} = [Ca^{2+}] [CO_3^{2-}]_{observed} / [Ca^{2+}] [CO_3^{2-}]_{equilibrium} \tag{3.2}$$

Clearly, D_{sat} equals 1 for saturation, less than 1 for undersaturation, and is greater than one for supersaturation.

Whenever D_{sat} is greater than one, we expect spontaneous precipitation. However, even though D_{sat} is indeed greater than one in the surface waters of tropical oceans, inorganic precipitation is negligible. It is commonly assumed that the presence of magnesium interferes with the expected reaction. Thus we need to find an unusually high D_{sat}, if we are to see inorganic precipitation. Additionally, the presence of appropriate crystal nuclei should be favorable.

High temperature, and a low CO_2 content in the water, increases the product $Ca^{2+}] [CO_3^{2-}]$ by increasing the concentration of carbonate ion. The effect of CO_2 is readily seen from the following equations:

$$CO_2 + H_2O = H^+ + HCO_3^- \tag{3.3}$$
$$HCO_3^- = H^+ + CO_3^= \tag{3.4}$$

Almost all the inorganic carbon in the ocean is in the form of bicarbonate, HCO_3^-. Removal of CO_2 drives the reaction of Eq. (3.3) to the left, subtracting hydrogen ions. In turn, this drives the reaction of Eq. (3.4) to the right, opposing the change in hydrogen ions. Increasing the temperature lowers the solubility of CO_2. Also, CO_2 is taken up by algae during photosynthesis — hence the precipitation of $CaCO_3$ as crusts on many shallow water tropical algae.

3.7.2 The Bahamas are ideally suited to test for inorganic precipitation. They are surrounded by some of the warmest, most alkaline waters in the ocean. There is hardly any terrigenous input, and pure carbonates are being precipitated. The Great Bahama Bank is extremely flat and usually less than 5 m deep (Fig. 3.8a). High rates of evaporation and low rates of rainfall result in increased salinities, which can attain over 40‰. Hence $[Ca^{2+}]$ (observed) is increased in Eq. (3.2). Heating of the water by the tropical sun further increases D_{sat}. Algae grow abundantly on the sea floor, removing CO_2 during the day. Thus conditions are indeed very favorable for the precipitation of carbonate.

Two types of calcareous particles on the sea floor of the Great Bank have been considered as possible products of inorganic precipitation: *aragonite needles* (length of a few micrometers) and *oolites* (diameter near one-third of a millimeter) (Fig. 3.8b).

The origin of the aragonite needles has been a puzzle for some time; both direct precipitation and mechanical break-up (perhaps aided by boring clams, deposit feeders etc.) of pre-existing skeletons has been suggested. However, modern investigations, using scanning electron microscopy and stable isotope analysis indicate that most of the needles are formed within certain algae.

The oolites are spherical objects formed of concentric layers of aragonite needles and organic matter. They are abundant on the outer rim of the Great Bank, in the shallowest water. Here in the zone of growth they are alternately moved about by strong tidal currents, or rest just below the sea floor, buried by

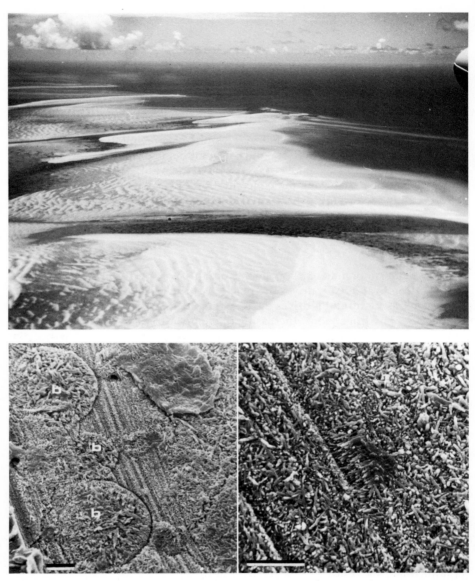

Fig. 3.8. Calcareous oolites, produced within tidal zone by algal activity. *Upper* air photo of oolite sand bars, Bahama Banks. Note tidal channels. (Photo courtesy D. L. Eicher) *Lower* SEM photos of ooids, scale bar 5 μm. *Left* slightly etched section. Three secondarily filled borings *(b)* intersect the concentric laminae of primary oolite coating. *Right* close-up of oolite laminae, showing acicular aragonite needles. (Photos D. Fütterer, Kiel)

other oolites. Recent laboratory experiments suggest that ooids are only formed if sufficient organic matter is present. According to high-resolution stereoscan investigations, precipitation is largely through biocalcification by unicellular algae. Hence carbonate precipitation directly from seawater seems to be negligible, under present-day conditions.

3.7.3 Dolomite. One might expect that dolomite [Ca, Mg $(CO_3)_2$], would precipitate from seawater because it is much less soluble than aragonite. There is, after all, plenty of magnesium in seawater (Table 3.1). However, such precipitation has not been observed in today's ocean. Dolomite apparently forms *within* calcareous sediments, either through partial replacement of Ca with Mg, in pre-formed carbonate, or by precipitation from pore waters, or both. A classic area for the study of this process is the southern margin of the Persian Gulf (Fig. 3.9). Here the intertidal flats lie behind barrier islands, within a lagoon with warm and very saline water. In the pore waters of sediments above low tide, magnesium

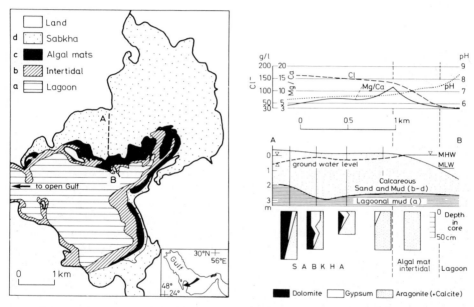

Fig. 3.9. Present-day dolomite formation. *Left* lagoon on the southern coast of the Persian Gulf *(a)*, floor covered with calcareous mud. Intertidal zone *(b)*, and algal mats of uppermost litoral *(c)* rim the lagoon. Intermittently flooded evaporite flats *(d,* sabkhas) follow.
Right geochemical profiles from sabkha to lagoon. Chloride ion concentration (normal seawater: 19 g/l) in the lagoon is 30 g/l, due to evaporation. Algal mats are exposed during low tides. Their interstitial waters reach 130 g/l. Gypsum forms here. The precipitation of calcium sulfate extracts Ca^{2+}, hence the rise of the Mg/Ca ratio to greater than 10 at the upper algal zone (for normal seawater, the ratio is 3). In the near-surface interstitial waters of the sabkha, evaporation is greatly enhanced by high temperatures (40° C and higher). Chlorinity reaches values of more than 150 g Cl^-/l — close to saturation with halite (table salt). Gypsum precipitates. The warm, Mg-enriched (and sulfate-depleted) brine sinks and reacts (at pH between 6 and 7) with the aragonitic subsurface sabkha sediment to produce dolomite. The resulting sediments are shown to the lower right. The dolomite crystals are small, less than 0.02 mm (silt). (After L. V. Illing et al., 1965, Soc Econ Paleontol Mineral Spec Publ 13: 89, simplified)

71

is greatly enriched with respect to calcium due to the precipitation of gypsum (calcium sulfate) in adjoining evaporite pans.

Obviously, in this instance of dolomite formation the dolomite is closely associated with evaporite flats and unusually warm and saline water — an environment unfriendly to higher organisms. Thus, we should not expect many macrofossils in sediments rich in this kind of dolomite. Dolomite also has been found in deep-sea sediments deposited under anaerobic conditions, both in the mid-Cretaceous of the Central Atlantic, and in Pleistocene organic-rich muds in the Gulf of California. Recent experiments on dolomite formation (P. A. Baker and M. Kastner, 1981, Science 213:214) suggest that the removal of the sulfate ion from the interstitial waters (by reduction and precipitation as iron sulfide) is the crucial step allowing dolomite formation to proceed. If so, the importance of the precipitation of gypsum in evaporite lagoons may lie in the decrease of the sulfate concentration rather than in the increase of the Mg/Ca ratio.

3.8 Hydrogenous Sediments

With the question of dolomite formation we have entered the evaporite environment. Within shelf and slope, the bulk of hydrogenous sediments are evaporitic salt. Strictly speaking, calcareous shells and skeletons also are hydrogenous, since they originate in the water. However, we have called the minerals precipitated by organisms biogenous, and set them apart.

3.8.1 Marine Evaporites are those sediments which form on evaporation of seawater. Restriction of exchange with the open ocean, in a semi-enclosed basin, is necessary to drive the salt content high enough for precipitation to begin. Such restricted bodies of water are (1) coastal lagoons; (2) salt seas on the shelves; or (3) early rift oceans in the deep sea. A special case is the Mediterranean, which was partially isolated in the latest Miocene, 6 to 5 million years ago.

How much salt can be produced by evaporating a 1000 m-high column of seawater? Salt constitutes 3.5% (or 35‰) of the weight of the column, its density is about 2.5 times that of water. Thus, we would obtain about 14 m of salt. Most of this would be table salt (halite) (see Table 3.1). The least soluble salts precipitate first: calcium carbonate (aragonite), and calcium sulfate (gypsum). To precipitate halite, the brine needs to be concentrated about tenfold. Many evaporites only contain carbonate and gypsum (or anhydrite), others have thick deposits of halite or, rarely, of the valuable (and very soluble) potassium salts. This sequence of mineral precipitation was experimentally established by Usiglio in 1849.

An evaporite basin, besides having restricted access to the open ocean, must lie within an *arid climate*. New seawater must be delivered from time to time. This seawater must be concentrated by evaporation. If only gypsum is to form, a concentration beyond three fold must be prevented, by new incursions of seawater, or else halite must be removed during or after each evaporation cycle (Fig. 3.10). If only halite is to form, left-over brine from a gypsum-precipitating basin

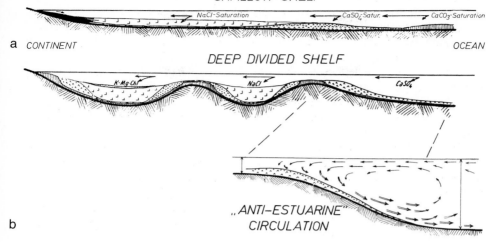

Fig. 3.10. Possible models for marine evaporite formation.
a Serial fractionation in very shallow and extended basins. Saturation of different salts is reached in a series ocean → land. Terrigenous particles may be supplied from land. Recent example: Adshi-darja Lagoon attached to the Caspian Sea by Kara Bogaz Inlet (chemical conditions there are not fully comparable with open sea).
b Serial fractionation and differential preservation in deeper basins divided by sills. Saturation of different salts is reached in a series shallow → deep water. *Detail,* only gypsum is precipitated near the sill. Halite saturation is not reached, because brine sinks down to the basin escaping further evaporation. Sill depths can be considerably reduced by carbonate and/or gypsum precipitation. No Recent example known. (After G. Richter-Bernburg, 1955, Dtsch Geol Ges 105 [4]: 59)

must be available, or the halite must be leached from elsewhere and brought into the basin by nonmarine waters. *Differential preservation,* then, and *serial fractionation* are the key processes in controlling the chemistry of salt deposits. In lagoonal settings a periodic covering of existing salts, by wind or flood deposits, may be necessary to prevent redissolution by the next invasion of ocean water.

3.8.2 Phosphorites. As is true for nonskeletal calcium carbonate, marine phosphorites are on the boundary line between hydrogenous and biogenous origin. After all, phosphorus is intimately associated with the life cycle on Earth, in the sea as well as on land. In fact, it is likely that the availability of phosphorus to photosynthetic organisms ultimately controls the fertility of the ocean, and hence also the formation of biogenous deposits.

Phosphorites deserve special attention for the reason of this tie-in to ocean fertility, but also because of their economic value (Sect. 10.3.1).

Marine phosphorites vary in composition. A general formula is $Ca_{10}(PO_4, CO_3)_6 F_{2-3}$, with increases in carbonate going parallel to increases in fluoride (or hydroxide). Modern phosphorites typically occur in areas of high productivity — off Southern and Baja California, off Peru, off South Africa, for example. They

73

occur as nodules up to head size and as irregularly shaped cakes. Commonly, and especially in the geologic record, they occur as replacement of previously deposited carbonates, or as mineralization of pre-existing organic matter. Off California these deposits contain about 25% to 30% P_2O_5 and 40% to 45 % CaO, but these values vary in other regions. The common depth of deposition is on the shelf and upper slope. Fossil phosphorites are abundant in places: in Florida and Georgia, Miocene phosphorites are mined on a large scale; in West Africa, Eocene deposits are being exploited. Seamounts bearing Cretaceous carbonates commonly carry phosphorites.

The association of geologically young phosphorite deposits with present-day regions of upwelling (Fig. 3.11) suggests that the source of the phosphorous is organic matter. Apparently, the algae growing in these regions, in the surface waters, extract the phosphorus from the water, and crustaceans and fish concentrate it further in their bodies and excrement. During decomposition of organic debris on the sea floor, much phosphate is released to the interstitial water (and also to seawater). Thus, interstitial waters right below the sea floor may become saturated with the phosphate mineral apatite. Precipitation of apatite, replacement of pre-existing carbonate mineral, and impregnation of sediment can then proceed. The resulting phosphatic concretions are resistant to transport by currents; they can be mechanically concentrated during periods of sediment reworking.

3.8.3 Iron Compounds are abundant in both oceanic margin sediments and in the deep sea since iron is one of the most abundant elements on Earth. On the slopes, the high supply of organic matter commonly leads to oxygen deficiency, and to sulfate reduction by marine bacteria, in the uppermost sediment layers. This process results in H_2S formation, and in the precipitation of *iron sulfide* (pyrite). In the deep sea, on the other hand, oxygen is generally plentiful, and essentially all iron occurs in its oxidized form, as *iron-oxide/hydroxide* (goethite). Here it is especially associated with manganese deposits (Chap. 10).

The reduction of sulfate in anaerobic sediments, and the associated precipitation of iron sulfide, is a geochemical process of major importance, which bears on the control of oxygen abundance in the Earth's atmosphere. Carbon can be deposited as carbonate or as organic carbon; sulfur as sulfate (gypsum) or as sulfide (pyrite); iron as iron oxide or as iron sulfide. In each case the second phase — which is favored by anaerobism in the ocean — liberates oxygen to the system. Thus, an increase in anaerobism leads to a decrease in oxygen consumption and to an increase in oxygen supply. This negative feedback mechanism helps stabilize the oxygen content of atmosphere and ocean.

An iron-mineral which has been much studied and discussed is *glauconite*. It is a greenish silicate common in shallow marine areas. Chemically it is a poorly crystallized mica, rich in potassium (7%–8%) and in iron (20%–25%). Geologists tend to denote as "glauconite" any small green earthy pellet recovered from the sand fraction of marine sediment. These pellets are commonly shaped like the interior of foraminifera or like fecal material, indicating their locus of growth. Apparently the association with decaying organic matter (fecal pellets,

74

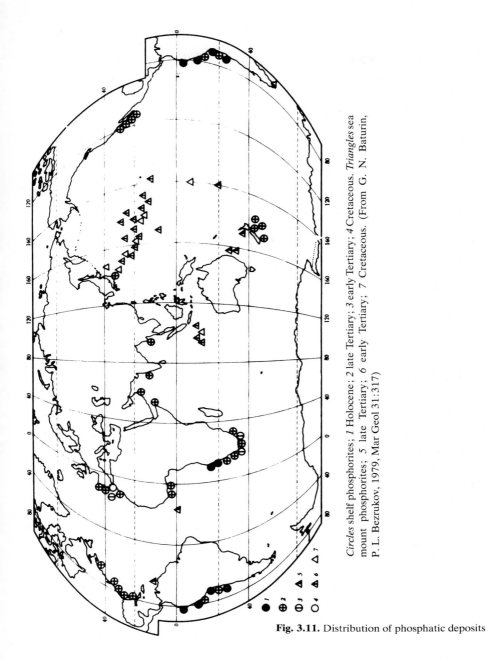

Circles shelf phosphorites; *1* Holocene; *2* late Tertiary; *3* early Tertiary; *4* Cretaceous. *Triangles* sea mount phosphorites; *5* late Tertiary; *6* early Tertiary; *7* Cretaceous. (From G. N. Baturin, P. L. Bezrukov, 1979, Mar Geol 31:317)

Fig. 3.11. Distribution of phosphatic deposits

interior of shells) is a necessary condition of growth: part of the iron in glauconite is reduced iron. A high concentration of iron in interstitial waters (at conditions intermediate between reduction of iron oxide and precipitation of sulfide) appears to be favorable for glauconite formation, as is the presence of the right kind of clay to convert into the glauconite mica.

From fossil marine sediments we know types of marine iron deposits which are not found today, such as the iron oolites abundant in the Jurassic formations of England, eastern France, and southern Germany. These "minette" iron ores of Europe have been mined for more than a century. The fact that their origin is still a mystery once again illustrates how little we know about the chemistry of ancient seas.

3.9 Sedimentation Rates

Consider a fossil reef section exposed in a mountain valley, or a sequence of layers of limestones or shales, telling about a certain period of Earth's history.

How long did it take to build that reef?

How much time is recorded in those layers? The *idea of geologic time*, which is so fundamental to geology, is quite young. Essentially, it starts with James Hutton (1726–1797), and its chief protagonists were Charles Lyell (1797–1875), and Charles Darwin (1809–1882).

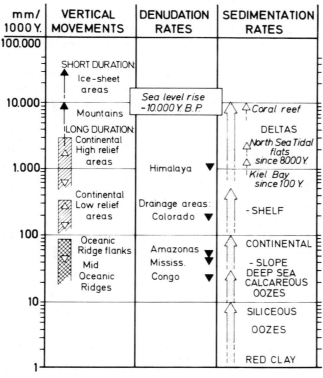

Fig. 3.12. Rates of vertical crustal motion, of denudation, and of sedimentation rates. Scale in mm/1000 years, or m/million years. Postglacial sea level rise for comparison: 100 m in 5000 yrs. Note interaction with recently deglaciated land *(arrows to left)* and with coral growth rates *(to right)*. (After E. Seibold, 1975, Naturwissenschaften 62: 62; modified)

However, before radioactivity was discovered (in 1896, by Marie Curie), and applied to the geologic record, there was no way of telling just how much geologic time might differ from the chronology which scholars had derived from the account in Genesis. Nevertheless, some early guesses proved remarkably close. Reade, in 1893, by observing present-day rates of denudation and of accumulation, estimated an age for sediments of a time span over 10,000 times greater than that admitted by Bishop James Ussher (1581–1656). J. G. Goodchild, in 1897, estimated 704 million years, a remarkably lucky guess!

A general overview of sedimentation rates is given in Fig. 3.12.

On the whole, high rates occur at the edges of the continents, especially in estuaries and marginal basins with river influx. The lowest sedimentation rates occur in abyssal regions far away from the continents. Characteristic values for continental slopes are 40 to 200 mm/1000 years, for the deep sea 1 to 20 mm/1000 years. Coral reefs build up at rates of 1 cm per year, that is, 10,000 mm/1000 years! (see Sect. 7.4).

The most reliable estimates of sedimentation rates are possible in the case of *annual layers* or *varves*. Counting of the varves in the Black Sea yields a rate near 400 mm/1000 years, and in a bay of the Adriatic Island Mljet, 250 mm/1000 years (Fig. 3.13). In the Santa Barbara Basin of California varve counts yield one millimeter per year. Unfortunately, such ideal calendar pages are extremely rare in the marine realm. They can only be preserved where burrowing organisms cannot live, that is, in the absence of free oxygen above the sediment-water interface.

It is somewhat hazardous to apply sedimentation rates found for a given type of sediment to a geologic sequence of such sediments. Episodic or periodic events, especially erosional events, can change overall rates considerably. In many cases

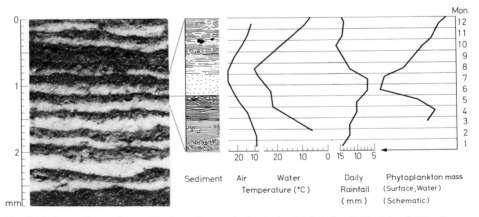

Fig. 3.13. Annual layers (varves) in the sediment of a bay in the Adriatic Sea (Mljet Island). The photo shows light and dark laminae. One light–dark pair corresponds to one year. The lower boundary of light layers are generally sharp; they are due to precipitation of carbonate by phytoplankton, which blooms in early summer (record to the right) as temperature rises. In fall and winter, rains bring terrigenous matter which — together with organic detritus — provides for dark colors. The interpretation of the varves is a complicated matter; recent studies use statistical procedures to reconstruct climatic conditions in detail. (Photo E. S.)

77

sediments of a certain depositional regime are underlain by erosional features of the same regime: ice-polished rocks under moraines, wind-polished pebble pavements under dunes, channels under river gravels, lopped-off sea floor beneath graded storm deposits (shallow water) or turbidites (deep water). Erosion, of course, means loss of information to the geologist. Missing sections — *hiatuses* — are erased history, lost time as it were, whose duration can hardly be estimated. On the other hand, the presence of hiatuses suggests we must look for information about the processes producing them. Also, hiatuses extending over large regions can be used for the correlation of sediment sequences, especially in seismic stratigraphy.

4 Effects of Waves and Currents

We have seen that the morphology of the ocean margins is largely determined by tectonics and sediment supply (Chap. 2), and we have reviewed the various types of sediments which are involved (Chap. 3). We now turn to the all-important role of water motion in determining the distribution of sediments on the sea floor. A first impression of this role can be gained from contemplating Fig. 4.1 which illustrates the redistribution of sediment supplied by a river in a setting typical for southern California. The water motions indicated are not the only ones that need to be considered, as we shall see.

Waves and currents leave their imprint on the sea floor in many ways, as depositional and erosional features. Familiar examples are wave forms on the sediment surface, from the smallest ripple marks to large submarine dune fields. Others are bedding structures within the sediment, from beach laminations to thick graded layers, and also the grain size of the sediment, from muddy lagoonal

Fig. 4.1. Redistribution of sediment on the continental margin, by water motion. The drawing reflects a West Coast geographic setting; note river input, longshore transport (powered by wave energy), interception by submarine canyon. Fine sediment bypasses the shelf, and comes to rest on slope or deeper. Note starved beaches and rocky shelf beyond canyon. (Based on a drawing in D. G. Moore, 1969, Geol Soc Am Spec Pap 107: 142)

deposits to the highly sorted sands on wave-washed beaches. Scour-marks, channels, and clean-swept banks and submarine plateaus are well-known products of erosion.

Just how effective are waves and currents as sculptors on the ocean floor? And what are the clues to be studied if we wish to reconstruct the wave and current regimes from the geologic record?

4.1 Sediment Transport

4.1.1 Role of Grain Size. The details of the relationships between water motion and sediment response are by no means clear, despite considerable study. One problem is the complicated feedback between water motion at the interface, and the changing character of the sediment surface. It is difficult to predict what will happen in one situation, from studying another, because grain size distributions, porosity, and cohesiveness of the sediment show large variability.

Perhaps the most basic question is, how strong does a current have to be to move sediment?

It is reasonable to expect that the coarser grains need more of a push than finer ones (Fig. 4.2). Pebbles with a diameter of 10 mm do not start to move until the average velocity of the current near the sea floor is near 2mls. Grains of 1 mm size move at 0.5 m/s. We can readily conclude, with a view to the geologic record, that fine grains will be moved more often than coarse ones in a downstream direction, as strong currents are less frequent than weak ones. Hence, grain size will normally decrease in the downstream (or down-current) direction. This is a first clue to direction of transport, a clue which can be applied along the coast, the shelf, or even in the deep sea.

However, this simple and obvious relationship between grain size and water velocity is valid only down to sizes of 0.1 to 0.2 mm. When grain sizes decrease below this value, water velocities may have to *increase* again to initiate erosion (Fig. 4.2). Why should this be so?

When settled on the floor, the very fine sediments tend to produce a smooth surface, which reduces turbulence at the interface and thus the opportunity for impact of fast water particles on the sediment particles. More importantly, the smaller the grains, the larger the surface area available for grain-to-grain cohesion, which can form a strong bond after compaction. Thus, fine-grained sediment resists erosion more than coarse sediment. Another conclusion follows: the sand grains between 0.1 and 0.2 mm are rather mobile on the sea floor; they are the nomads among the sediment grains. They move easily, at current velocities of only a little over 0.3 m/s. They are the sands to travel farthest, and they are commonly found, therefore, on intertidal flats far removed from the source.

4.1.2 Role of Velocity. How exactly does water move the grains? The velocity of the current decreases toward zero at the interface itself. Hence the values of

80

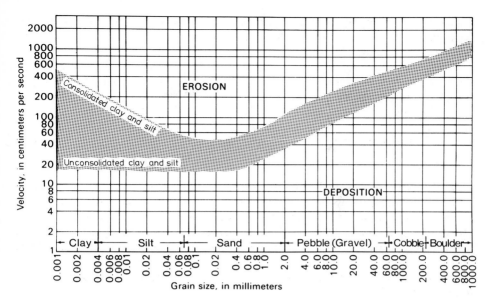

Fig. 4.2. Diagram relating current velocity somewhat above the bottom to the size of particles of a given class which it erodes ("Hjulstrøm Curve"). Graph applies to well-sorted sediment only. (After A. Sundborg, 1956, Geogr Ann 38: 127, in J. Gilluly et al., 1968, Principles of geology, W. H. Freeman, San Francisco)

current velocity given are valid for some distance *above* the floor. Exactly how much effect such a current has near the interface depends on the roughness of this surface and on the turbulence this creates. The turbulence leads to sudden changes in the impact of water on a grain sticking out at the surface. As the current velocity increases, the frequency and force of impact pulses increases, and some grains start to move. This leads to the impacting of grains by other grains and soon more and more grains start to roll and jump over the floor. The rolling and jumping grains are the *bed load* of the current (Fig. 4.3).

If the velocity of a current increases further, turbulence increases and the jumps become higher, with more grains joining. For any one grain the contact with the floor decreases, and we now have a considerable *suspended load* within the water. Since fine grains settle more slowly they spend more time in suspension than coarse ones. Thus, there is a statistical distribution between bed load and suspended load, corresponding to grain size distributions and turbulence.

The grains traveling in the bed load, naturally, will impact each other more than those in suspension. We can conclude that there is another clue to direction of transport: signs of abrasion on bedload particles should increase down-current. The *degree of rounding* of pebbles, for example, is such a clue. Sand grains, too, may bear this information, although it is assumed that abrasion is much slower than for pebbles — 300 to 400 times less effective. Below a size of 0.25 mm, however, rounding and length of transport have no simple relationship. In fact, rounding may *decrease* down-current as the sands become finer and hence more irregular.

81

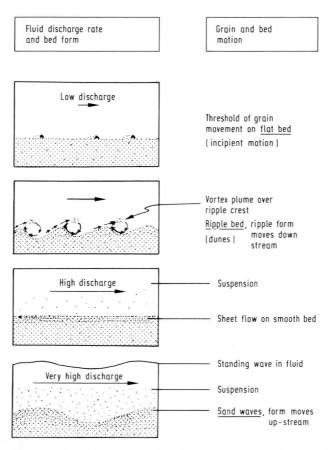

Fluid discharge rate and bed form	Grain and bed motion

Low discharge →

Threshold of grain movement on flat bed (incipient motion)

Vortex plume over ripple crest

Ripple bed, ripple form (dunes) moves down stream

High discharge →

Suspension

Sheet flow on smooth bed

Very high discharge →

Standing wave in fluid

Suspension

Sand waves, form moves up-stream

Fig. 4.3. Schematic representation of change in grain motions as velocity of water increases (flume experiments). (After D. L. Inman, in F. P. Shepard, 1963, Submarine geology, 2nd ed. Harper and Row, New York)

What if the current velocity of a sediment-carrying water body decreases from some high value? The coarse material settles out first, and then the finer particles in regular succession (Fig. 4.4). The values for current velocities at which the sediment comes to rest are *lower* (by about 30%) than the ones for erosion: it is easier to keep sediment moving than to set it in motion from rest. For the suspended load, of course, there ist no velocity minimum for transport, as in erosion, since *settling* now dominates over interface conditions. One conclusion is that any current activity will tend to separate clay-size from sand-size material, since the presence of sand facilitates erosion, and the clay stays in suspension long after the sand settles back out.

4.1.3 Role of Events. Only short events producing suspension are needed to allow considerable transport: once the sediment is brought into suspension, by

82

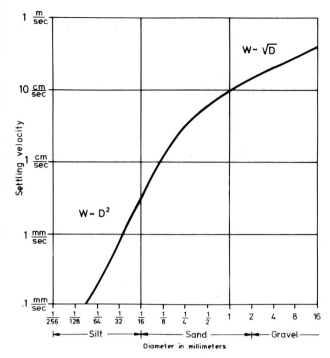

Fig. 4.4. Settling velocities of quartz grains in water. Note that silt and fine sand settles according to $W \backsim D^2$, but coarse sand and gravel according to $W \backsim \sqrt{D}$. ($W \backsim D^2$ is known as Stokes' Law.) (W is settling rate.) (After W. W. Rubey, 1933, Am J Sci 25: 325, in C. O. Dunbar, J. Rodgers, 1957, Principles of stratigraphy. John Wiley, New York)

currents, waves or animals on the sea floor, even a relatively weak current can keep it moving. The occurrence of suitable events may be separated by long time intervals; nevertheless, the sediment will move. Thus, in describing the role of environmental factors in producing sediment transport there is a problem of *scale,* that is, time-scale. The coastal engineer knows that storms do much of the moving of sediment — and damage — along the shores. Yet, his measurements may only include a few years' worth of changing conditions. The marine geologist, thinking in millions of years, is faced with a record which may contain a large proportion of unusual events, and which is hard to compare with the results of laboratory experiments, or even field experiments.

Experiments on sediment transport, for example in artificial waterways (flumes), have an additional scale problem, the scale of size. Small-scale experiments do not necessarily reproduce the large-scale phenomena of nature.

83

4.2 Effects of Waves

4.2.1 Waves and Offshore Sediment. Sooner or later every marine geologist becomes familiar with some of the less pleasant effects of waves, when at sea. Here we are concerned with the effects on the sea floor, however. Waves consist of near-circular motions of water particles; the upper part of the circle moves with the wave, the lower part against it (Fig. 4.5).

The motion quickly decays with depth. At depths greater than one half of the wave length there is virtually no motion. Thus, surface waves only involve the uppermost part of the water column. Internal waves also exist at the thermocline, and at other density discontinuities in the upper few 100 m. However, their effects are poorly known.

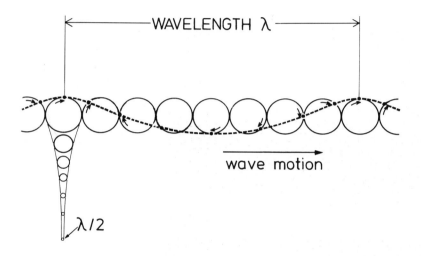

a

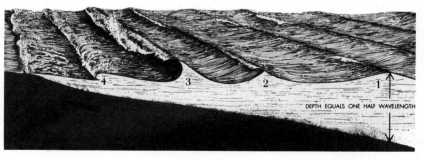

b

Fig. 4.5a, b. Wave motion. **a** Sinusoidal wave, particles describe circles. **b** Formation of breakers. The open ocean wave approaches the beach *(1)*, "feeling bottom", it slows and steepens *(2)*. It becomes unstable and breaks *(3)*, producing a body of foaming water which runs across the beach *(4)* transporting sediment. (After W. Bascom, 1959, Sci Am 201 (2) 14 and 1960, Sci Am 203 (2) 80 modified)

A surface wave running up onto the shore from deep water "feels bottom" when the depth of water becomes less than one fourth of the wave length. At this point, the circular motion of the water particles becomes more and more elliptical, and there is a back-and-forth motion directly on the sea floor. In general, the maximum depth to which sand is moved, the "wave base", is near 10 to 20 m. In exceptionally strong storms, wave motions can reach considerably deeper. Ripple marks are seen on shelves even out to the shelf break. However, the relative importance of surface waves, internal waves, tides, and currents in producing such ripples is not always clear. Recently symmetrical ripples were discovered in fine sands of the outer shelf off Oregon, at depths to 200 m, with crest-to-crest lengths of 10 to 20 cm, and with the crests parallel to the coast. These oscillation ripples are thought to be produced by winter storms on the exposed shelf. The measurement of ripples (Fig. 4.6) yields clues as to the water motion which produced them.

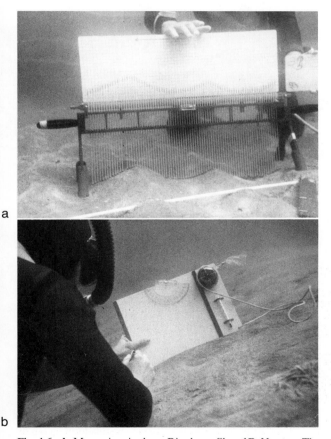

a

b

Fig. 4.6a, b. Measuring ripples. **a** Ripple profiler of R. Newton. The vertically movable rods reproduce the ripple topography on a grid. **b** Diver measuring the inclination of a lee slope of a giant ripple, north of Fehmarn Island, Baltic Sea (see Fig. 4.7). (Photos Diving Group, Geol. Inst. Kiel)

The nature of the waves working the sea floor helps determine not only the distribution of various types of ripples, but also the character of the sediment. Above wave base, fine sediment is put into suspension periodically, and is transported away by currents. It settles in quiet environments at greater depths. Below wave base, mud can accumulate. The composition and the rate of production of bottom-living organisms changes greatly across the wave base, presumably largely because of the change in sediment character.

The rule "coarse sediment equals shallow water, fine sediment equals deep water" is strikingly illustrated in the western Baltic Sea (Fig. 4.7). Some channels

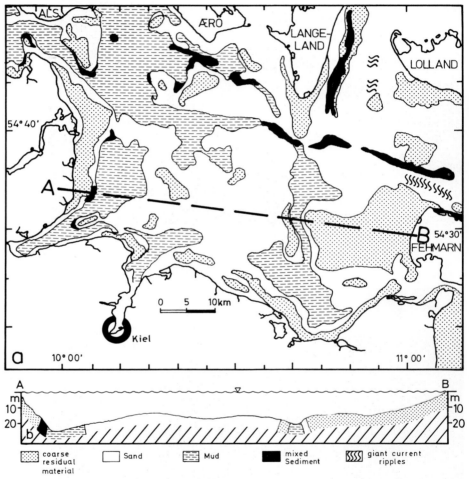

Fig. 4.7 a, b. Sediment distribution in Kiel Bay: **a** map, **b** profile *A-B*. Sediment is mainly coarse to fine terrigenous material, the facies follows depth zones. Coarse residual deposits form on top of morainal debris, washed out by waves in shallow areas. Mud collects in basins. Mixed coarse and fine sediments are of complex origin, associated with strong currents; locally they contain manganese nodules in the western part

are characterized by a mixture of coarse and fine sediment. Note that the less exposed bays also tend to collect fine material, while coarse sediment characterizes the headlands.

4.2.2 Beach Processes. Nowhere the effects of water motion are more obvious than on the beach (Figs. 4.8 to 4.10). Each uprushing wave moves the sand, leaves

a

b

Fig. 4.8a, b. Beaches. **a** Hawaiian beach made of coarse shell sand (note steepness of profile). Breakers form at reef edge, well offshore. *Arrow* marks "swashline". **b** Pebble beach in East Anglia. Note the distinct berm crest. (Photos W. H. B)

a

b

Fig. 4.9a, b. Boomer Beach, La Jolla, California. **a** Exposed wavecut terrace with boulder beach. Sand has moved offshore, and also to southern end of embayment. This is the condition after storms (usually in winter). **b** Sand covers the terrace and much of the boulders. Summer condition. (Photos W. H. B)

swash marks at its upper limit and V-shaped backwash rills upon retreat. The balance between erosion and deposition of beach materials produces a typical profile, featuring a *berm* and a *foreshore* (Figs. 4.8b and 4.10). The balance

88

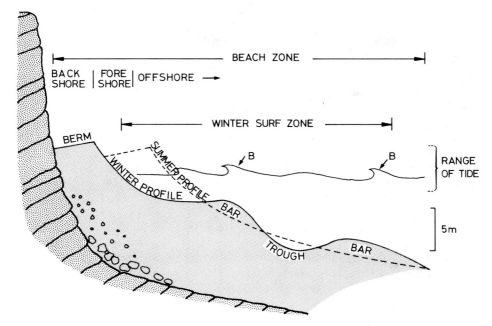

Fig. 4.10. Beach profile (southern California). The summer profile shows the build-up of the berm at the expense of the bars. During winter the surf removes sand from the berm, and the bars reform. Note the breakers over the bars *(arrows marked B)*. (After W. Bascom, 1960, Sci Am *203* (2) 80; modified)

between erosion and deposition shifts through time. After storms, in many cases, a considerable loss of beach sand can be noticed. The sand is eroded and moves offshore.

How do the beaches hold on to their sand? The answer to this question is that normal, more gentle wave action moves the sand back inshore. On the beaches of California, and elsewhere on many coasts, the power of wave action changes seasonally. Thus, winter beaches tend to be lower than summer beaches, and the out-going sand leaves rocks and gravel exposed where present (Fig. 4.9). The sand which collects offshore forms bars. The position of the bars can be recognized from the shore by observing the breakers which form on top of them (Fig. 4.10).

The waves which run toward the shore steepen up as they slow in shallow water, and at some point they break, spilling water down the front (Fig. 4.5b). This breaking occurs where the water depth is roughly 1.5 times the wave height (from trough to crest). Waves transport some water in the direction of travel, and this transport greatly increases during breaking. Thus, water piles up on the beach, or on the *foreshore,* to be exact (Fig. 4.10). In many places, the piled-up water returns to the ocean in narrow streams, the *rip currents* feared by inexperienced ocean swimmers. These currents can be fast (1 to 2 m/sec) and can carry swimmers beyond the breaker zone. Here the currents disperse. They also can carry sediment, and produce channels and ripple marks within the breaker zone.

Water transported by waves also causes *longshore currents,* which can move sediment parallel to the beach. Since the breakers are very efficient in suspending the sediment, the longshore currents (velocities up to 1 m/s) can move a large amount of material (Fig. 4.11a). On both the U.S. East and West Coasts the longshore drift is predominantly from North to South. On the shores of Southern California, sand brought by rivers moves south till it is intercepted by the head of a submarine canyon. Thus, south of submarine canyons, beaches tend to be starved for sand (Fig. 4.11b). The submarine canyon acts as a funnel, moving beach sediment as well as kelp and other debris down to the deep sea floor (Fig. 4.12).

Where beaches are missing or narrow, storm breakers can hit the coast with full force. Armed with gravel or sand, breakers can dig a deep notch into cliffs, even if these are made of hard rock. When the notches and sea caves become large and deep enough, the overlying material collapses and cliff erosion results. By grinding up the fallen rocks, waves produce additional beach sand (Fig. 4.13).

All over the world one commonly finds protective structures at the foot of cliffs. People are more mobile now than ever, they flock to the sea, they enjoy the

a   b

Fig. 4.11a, b. Longshore transport of beach sand. **a** Venice Beach, Los Angeles, looking north. Note pile-up of beach sand on northern side of large jetty at bottom. **b** La Jolla Point, San Diego, looking south. Sand moves in from the north; beach ends abruptly before reaching promontory. Sand moves offshore in La Jolla Canyon *(arrow).* Dark fields north of point are kelp beds. Note starved pocket beaches on the point. (Photos W. H. B.)

90

Fig. 4.12. Sandfall in San Lucas Canyon, Baja California. The fall is about 10 m high. The photo was taken by Conrad Limbaugh, at a depth of about 55 m, in natural light (see F. P. Shepard, 1963, Submarine geology, 2nd ed. Harper and Row, New York, Fig. 148. Photo courtesy S.I.O.)

view atop the cliffs, and they tend to be unfamiliar with the realities of coastal erosion. Coastal engineering — which largely progresses by avoiding previous errors — has evolved considerably in areas where protection from storm waves has been part of the scenery for centuries, for example around the North Sea. Walls proved to be vulnerable to waves. The waves remove the support at the base, so that the wall topples over toward the sea. A more effective way to tame the sea is to pile up "rip-rap" (boulders) or "tetrapods", or to construct dams with gentle slopes, where the breakers can spend their energy gradually rather than all at once (Fig. 4.14). Energy-consuming friction and turbulence is increased by making rough surfaces on these slopes. The words of Francis Bacon (1561–1626) apply: "Who would rule Nature, must first obey her."

Fig. 4.13. Sediment from cliff erosion, Encinitas, California. The pebbles act as tools for undercutting of the cliff. Note fallen cliff rocks which are quickly ground up into small particles. (Photo W. H. B.)

Fig. 4.14a–c. Defense against surf action. Sylt Island, German North Sea. **a** Power of storm surf (Feb. 18, 1962) demonstrated by damage to beach wall and transport of heavy tetrapods. **b** Gently sloping wall breaks the power of storm tide breakers (fall 1961). **c** Interlocking tetrapods serve to dissipate wave energy before it hits the wall. (Photos E. S. **a** and J. Newig **b, c**)

92

4.3 Effects of Currents

4.3.1 Surface Currents. Of all ocean currents the *Gulf Stream* is perhaps the most familiar (Fig. 4.15a). It is the largest and most important current of the Northern Hemisphere, and is mighty indeed, transporting nearly one hundred million cubic meters per second (100×10^6 m^3). For comparison, extraordinary peak floods of the Mississippi might carry as much as fifty thousand cubic meters per second (50×10^3 m^3/s) — on half of a thousandth of the Gulf Stream transport.

The Gulf Stream, of course, is part of the great North Altantic gyre, complete with west wind drift in the north, trade wind currents in the south, and an eastern boundary current (the Canary Current) connecting the two. The center of the gyre is near 30° N, in the Sargasso Sea. There are five such gyres on the globe: North and South Atlantic, North and South Pacific, and Southern Indian Ocean. In each case the west winds and the tradewinds are the driving force, and the western and eastern boundary currents complete the circle (Fig. 4.15b, c). Normally, the currents of the open ocean reach down to about 100 to 200 m, and their velocities are low: a fraction of a knot (a *knot,* a nautical mile per hour, is close to 0.5 m/s). The Gulf Stream, however, and other fast narrow boundary currents (e.g., Kuro Shio off Japan) reach to a depth of some 1000 m and have velocities of a couple of knots or so ($\sim$100 cm/s). The Florida Current which issues from the Gulf through the narrow Florida Straits reaches velocities of 6 knots (300 cm/s).

The general outlines of the gyral circulation have been known for some time. Present research focuses on the meanders within currents like the Gulf Stream, on the eddies separating from such a current and on the mechanisms of mixing with the surrounding ocean.

The surface currents have profound geologic effects. They strongly influence weather and climate, by transport of heat and moisture. They also write the record of climatic changes on the sea floor, by controlling the production of biogenous deposits. Planktonic organisms, which mostly live in the upper few hundred meters, can actually be used as *tracers* of currents, much like drift bottles. Those planktonic organisms which form hard parts, of course, can trace out the path of surface currents on the sea floor, for the geologic record.

Currents, on the whole, run parallel to isotherms, and are strongest where temperature gradients are strongest. The reason is that temperature distributions are largely congruent to density distributions, which in turn are generally in equilibrium with currents. From these rules, ancient currents can be mapped from a reconstruction of the temperature field, based on plankton remains (see Sects. 7.2.1 and 9.2.2).

Surface currents and associated temperature distributions not only control plankton patterns, but they also control the growth of benthic assemblages, especially coral reefs (see Sect. 7.4). The western boundary currents bring warm water to high latitudes, the eastern ones bring cool water to low latitudes. Thus, the belt of tropical coral reefs is much wider in the west than in the east, within each ocean basin.

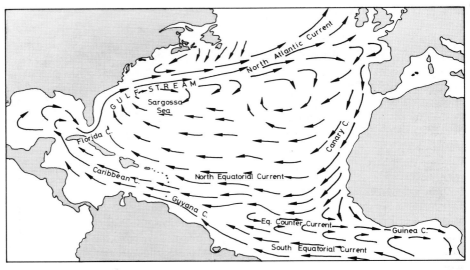

a

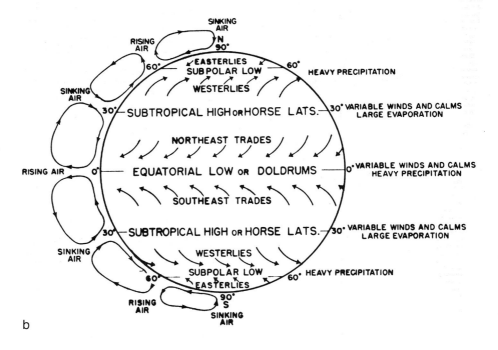

b

Fig. 4.15a–c. Gulf Stream, and wind-driven ocean circulation. **a** The Gulf Stream carries warm surface waters toward the seas of NW Europe. It is part of the giant subtropical gyre of the North Atlantic whose center ist the Sargasso Sea. (Mainly after G. Neumann, W. J. Pierson, 1966, Principles of physical oceanography. Prentice-Hall, Englewood Cliffs)
b The global wind system, schematically. (From R. H. Fleming, 1957, Geol Soc Am Mem 67: 87).

94

WINDS   CURRENTS

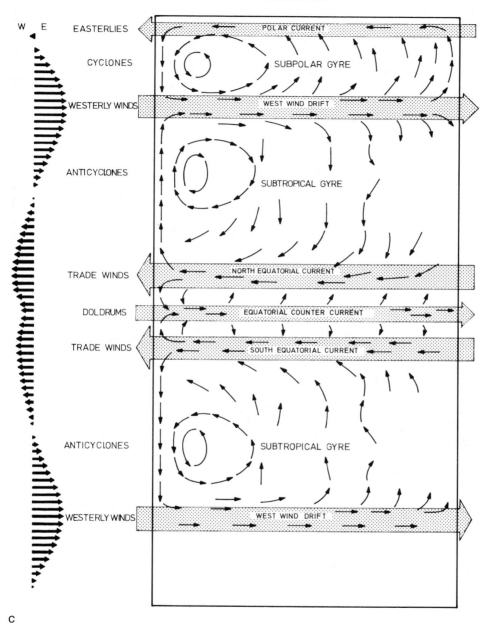

c Idealized rectangular ocean with the currents which would form as a result of the wind forces shown on the left. The asymmetry of the gyres is caused by the Earth's rotation. (From W. Munk, 1955, Sci. Am *193* (3) 96; modified)

The drift of ice-bergs and hence the transport paths of their load of rocks is controlled by currents, as are the paths of various other drifting materials such as trees — some of which are thought responsible for the dispersion of land animals on islands. The drifting larvae of benthic organisms *(mero-plankton)* reach their uncertain destinations with the aid of currents, and disperse throughout the ocean basins by island-hopping if necessary.

Finally, surface currents can affect the sea floor directly, through erosion on shelves and even on upper slopes. The deep-reaching Gulf Stream sweeps fine material off the Blake Plateau (see Fig. 1.3), keeping the manganese pavements there clean.

4.3.2 Current Markers. How can we detect the action of currents by studying the sea floor? The winnowing of fines from remaining coarser material, that is the "sorting" of sediment, was discussed earlier (Sect. 4.1.2). Usually it is necessary to make grain size determinations to shwo that sorting occurred. However, sometimes effects of winnowing can be seen in bottom photography, which show pavements of cobbles, nodules, shells, or other winnowed layers of coarse residual material. Also, scour marks behind obstacles are common indicators of currents.

The use of side-scan sonar, on the shelf and more recently in the deep sea, has greatly increased our knowledge about the activity of bottom-near currents, from their effects on the sediment patterns and the morphology of the sea floor (Fig. 4.16). Streaks of coarse material in fine sediments, indicating current action, were discovered in the entrance to the Baltic Sea by such acoustic sensing (Fig. 4.17).

The *direction* of a current can be readily deduced from photography and acoustic scanning. Current *strength* is more difficult to deduce. Direct observations on the effects of current strength are possible in tidal flats. The German and Dutch *Wadden* flats of the North Sea have long been a natural laboratory for these kinds of observations (Fig. 4.18).

We have seen some features which *parallel* currents. There are those that are *at right angles* also: the current ripples. Much like wind dunes, ripples have a gentle up-current side and a steep down-current slope (Fig. 4.19). This type of ripple differs from the oscillation type mentioned earlier (Sect. 4.2.1). The internal laminae of the current ripples consist of the down-current slopes made by material falling over the edge at the crest. Thus, fossil ripple marks *(cross-bedding)* yield clues to ancient current systems in contact with the sea floor.

The relationship between current velocities and ripple formation has been studied in the laboratory, in some detail. Ripple characteristics depend both on the type of sediment present, and on velocity distributions in a complex manner. The formation of ripples of dimensions of centimeters to decimeters starts at current velocities of about 25 to 100 cm/s. The origin and architecture of *giant ripples* and *underwater dunes* is less well known. These features may be several meters high, and tens, even hundreds, of meters apart. Some resemble the barchan dunes of the great deserts. What kind of currents are necessary to

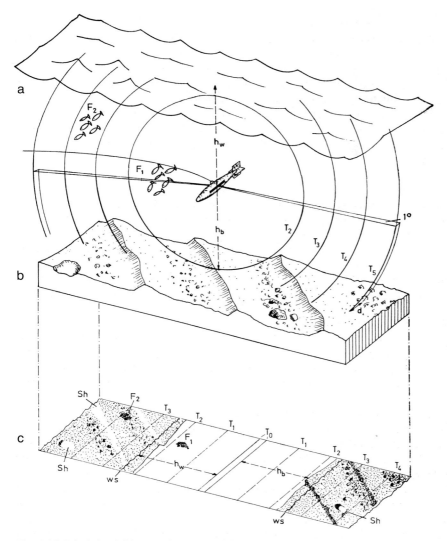

Fig. 4.16. Principle of side-scanning echo sounder.
a water surface; **b** sea floor with ripples, rocks, and a small depression *(left)*; **c** a section of the acoustic recording (sonograph) T_o outgoing pulse of the towed submerged sound source (the "fish"); T_1, T_2, etc., time markers (=distance to sound reflecting object). *Sh* acoustic shadow. F_1, F_2 schools of fish, and their acoustic image, *d* width of insonification on the sea floor (=resolving power). h_w distance "fish" to water surface *(WS)*; h_b distance "fish" to sea floor. (From R. S. Newton et al., 1973, Meteor Forschungsergeb. Reihe C 15; 55)

produce such dunes? Are the dunes relicts from periods of unusual activity on the sea floor? These questions are still open at the present time.

In coastal areas with a sufficiently large tidal range, alternation between high and low tide produces *tidal currents*. Considerable erosion can take place where

97

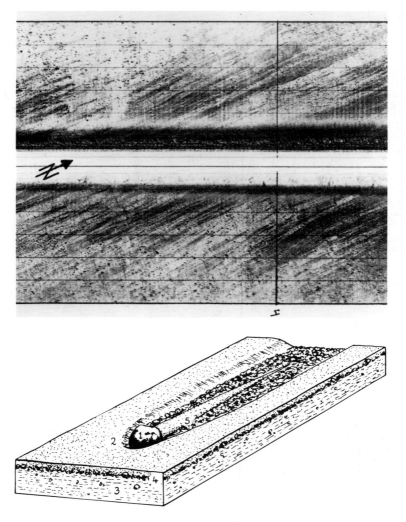

Fig. 4.17. Comet marks in the channel Store Belt, Baltic Sea, between the Danish islands of Langeland and Lolland. *Upper:* Sonograph of the sea floor. Length of record about 2 km, width ⁓150 m, water depth about 12 m. *Dark spots* are boulders (morainal debris). Some boulders have a "tail", toward left and down. These "comet marks" indicate currents from the north. (Sonograph by F. Werner, Kiel.) *Lower:* Morphology of a typical comet mark, as seen by divers. *1* Obstruction (boulder); *2* crescent scour; *3* till; *4* residual pebble layer (fossil erosional pavement); *5* "comet tail" of fine sand, built into erosional depression produced by turbulence behind obstacle. (After F. Werner et al., 1980, Sediment Geol 26: 233)

tidal currents are confined because of the high velocities attained and the large volumes of water which are commonly involved. In a number of harbors in estuaries, the tidal currents help flush out the entrance and prevent it from silting up. The inlets of lagoons on the East Coast are kept open by fast tidal currents. As these currents slow on either side of the inlet, *tidal deltas* form (see Fig. 3.8).

a   b

Fig. 4.18a, b. Current action in tidal flats. **a** *Mya arenaria* in life position uncovered by erosion of about 20–30 cm of mud, by lateral migration of tidal channel. Note the residual bivalve shells which collect as a "pavement" on the erosional surface. **b** Large current-produced ripples, covered by small ripple marks, German Bight north of Weser estuary. Ripples migrate ocean-ward *(to the left)*, hence they are produced by currents of the outgoing tide. (Photos E. S.)

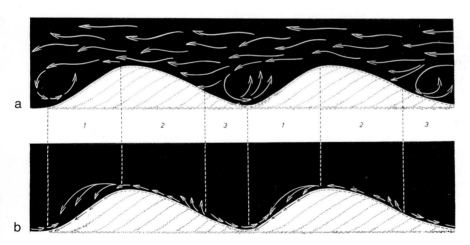

Fig. 4.19a, b. Water **a** and sand **b** movement at current ripple marks. Sand is eroded at the luff side *(sector 2)* and migrates to the ripple crest. From there sand is transported on the bottom *(black arrows)* or in suspension *(white arrows)* downward forming lee laminae. Horizontal bottom vortex in the ripple valley removes fine grains, concentrates coarse material, including shells. Therefore ripples move over coarser-grained base laminae. (After H. E. Reineck, 1961, Senckenbergiana Lethaea, 42: 51)

99

4.3.3 Upwelling. So far we have contemplated the effects of horizontal currents on the sea floor. *Vertical water motion* also affects the sea floor, in more than one way. The *overall mixing rate* of the ocean is ultimately tied to vertical motion, and this rate is closely tied to the productivity of the ocean. In turn, the sea's productivity influences what types of biogenous sediments (carbonate, silica, phosphates) will end up on what part of the sea floor.

A more obvious example for the effects of vertical motion is *upwelling.* Off coasts with eastern boundary currents, surface water has a tendency to move seaward, due to deflection by the *Coriolis Force* which results from the rotation of the Earth. Deflection is to the right on the northern hemisphere, and to the left on the southern one (Fig. 4.20a). The seaward motion is reinforced when winds blow off-shore. Outward-bound surface waters are replaced by cold, nutrient-rich waters from within the thermocline (100 to 200 m depth) (see Fig. 4.20b). Hence the low water temperatures off Northwest and Southwest Africa, California, Peru, and Chile, and hence also the high production of algal plankton, which feeds zooplankton, fish, and even birds up the food chain. The sediments below areas with strong upwelling are typically rich in organic matter. For example, sediments in the Walvis Bay area, Southwest Africa, have up to 20% of organic carbon (C_{org}). Also, usually opal is abundant: off Walvis Bay one finds up to 70% opal from the frustules of diatoms. Fish debris and other vertebrate remains also are increased in abundance, presumably delivering part of the phosphate necessary for phosphorite formation (see Fig. 3.11).

The sediment contains many detailed clues to the intensity of upwelling. The planktonic species (foraminifera, diatoms) tend to indicate cold water. The high supply of organic matter depresses the oxygen content, due to decay in the deep water and on the sea floor. In extreme cases anaerobic sediments with annual layers *(varves)* may develop. Using the clues described, it is possible to follow changes in position and strength of upwelling, and hence the migration and change of nature of the subtropic climatic belts in the geological record. This is a field of active research.

Upwelling is not restricted to coastal areas. The divergence of surface waters anywhere in the open ocean causes replacement from below. The great equatorial divergences in the Pacific and Atlantic Ocenas cause the well-known *equatorial upwelling* which we shall encounter when studying carbonate distributions (Chap. 8). The high productivity along these belts not only raises the sedimentation rates of calcareous and siliceous ooze, but also may be ultimately responsible — through concentration of trace metals in organisms and transport to the sea floor in organic debris — for the high values of copper, nickel, zinc, and other metals in Pacific manganese nodules (Chap. 10).

4.3.4 Deep Ocean Currents. There was a time, not so long ago, when the deep ocean was thought to be a quiet, calm environment whose currents are weak and relatively unimportant as far as shaping the sea floor. The first indication that this ist not necessarily so came from the calculations of Georg Wüst (1890–1977) and of Albert Defant (1884–1974) in the 1930's. They showed that the temperature-salinity distributions seen in the closely spaced profiles of the Meteor-Expedi-

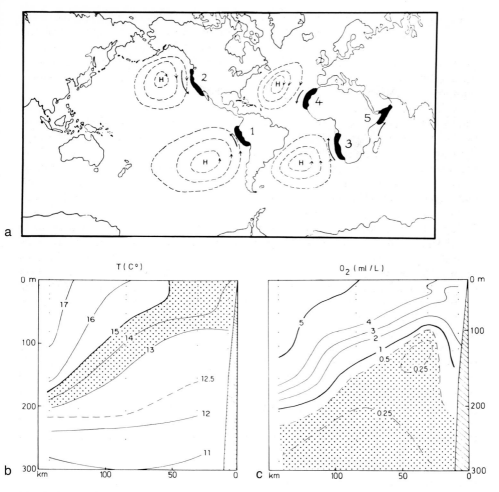

Fig. 4.20. Upwelling currents. *Upper:* Global distribution of main upwelling areas. Note position on the east side of subtropical gyres, at edge of »eastern boundary currents". Surface water moves offshore due to Coriolis Force, and is replaced from below. (From Science, 1980, 208: 39). *Lower:* Temperature and oxygen distributions in an upwelling area off San Juan, Peru, Sept. 1968 (S. Zuta et al., in R. Boje, M. Tomczak, 1978, Upwelling ecosystems. Springer, Heidelberg). Note the water layer near 200 m depth in the open ocean extends to the surface at the continent. This water is oxygen-poor and nutrient-rich. The high nutrient supply stimulates algal production (that is, growth of dinoflagellates and diatoms)

tion in the central Atlantic implied strong bottom-near flows driven by density differences (Fig. 4.21). Such currents hug the slopes of ocean margins and the flanks of mid-ocean ridges, and they follow density-surfaces which are practically horizontal. Hence, the currents are parallel to bottom contours, and marine geologists refer to them as *contour currents*. For physical oceanographers they are *deep geostrophic currents*. Their effects are clearly visible on deep sea

101

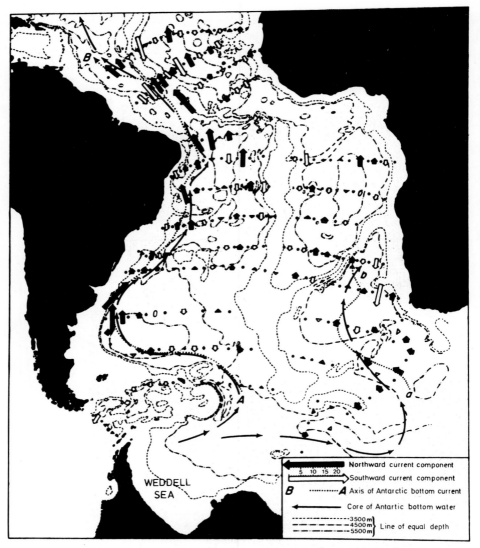

Fig. 4.21. Strengths of bottom currents in the South Atlantic, according to calculations of G. Wüst, based on density distributions. (After A. Defant, 1961, Physical oceanography, Pergamon Press, London)

photographs (Fig. 4.22). Such currents can erode, especially where confined to passages. Erosion occurs, for example, where the abyssal Antarctic Bottom Water passes through the Vema Channel off Argentina, or through the Samoan Passage, in the eastern South Pacific.

The *Antarctic Bottom Water* (AABW) fills the deepest parts of essentially all major ocean basins, through such passages. In the western South Atlantic and in the South Pacific, the AABW causes dissolution of carbonate on the sea floor —

a process that profoundly affects sedimentation patterns and the nature of the substrate for benthic organisms. The AABW originates on the shelf of the Antarctic, mainly in the Weddell Sea, by cooling of polar waters with admixtures of saline *North Atlantic Deep Water* (NADW). With a salinity near 34.7‰, when

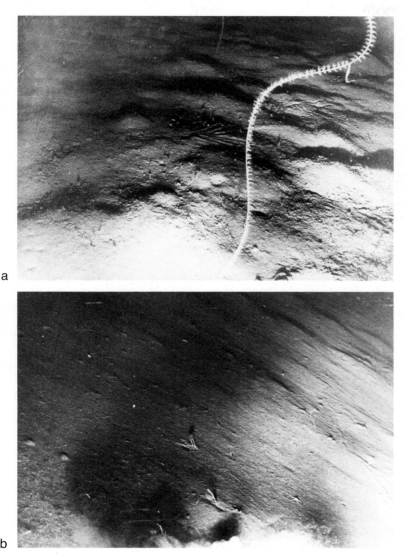

Fig. 4.22a, b. Effects of deep geostrophic currents. **a** Continental rise off the East Coast, 2.5 km depth. Little or no evidence of current action. (Organism is a "sea whip", pennatulid coral). **b** Same area, but 5 km depth. Current action is obvious from scour marks and lineations. Current flow is *from lower right to upper left,* about 20–30 cm/s. (Photos courtesy C. D. Hollister; see A. H. Bouma, C. D. Hollister, 1973, SEPM Pacific Section, Short Course, Turbidites and deep water sedimentation, Anaheim)

reaching a temperature of near $-0.4°$ C, the Antarctic surface water becomes heavier than the underlying water and sinks through it. Since it is also heavier than almost any water in the deep ocean, it sinks to the abyss as AABW and fills it from below. Its spreading depends on a deep access to a basin. For example, the Angola Basin can only be reached through the Romanche Deep, at the Equator. Hence the AABW makes a long detour to get there.

Besides chemical and mechanical erosion, AABW can produce small-scale features such as scour and ripple marks, and (presumably) large-scale ones, such as the giant dunes in the Argentine Basin.

The deep-towed side-looking sonar developed by F. Spiess and co-workers has revealed an abundance of dunes, sediment ridges, and erosional ravines in many places, even where deep currents had not been suspected (Fig. 4.23).

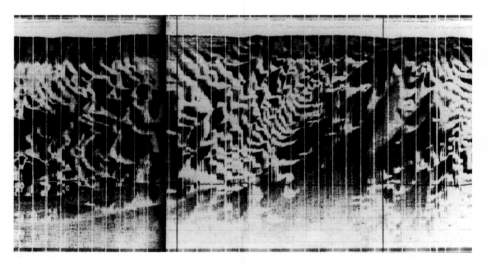

Fig. 4.23. Giant Dunes in Carnegie Ridge area, eastern tropical Pacific. Side-scan sonograph taken at 2.4 km depth, covering 1600×450 m. Sediment is calcareous ooze. (Photo courtesy P. Londsdale and B. T. Malfait, S.I.O.)

Of course, the possibility always exists that (a) calculations on current distributions are in error, and (b) the features seen have nothing to do with the present water motion, but are the witnesses of conditions long past.

The NADW sinks in the Norwegian Sea and falls over shallow passages over the Greenland–Faroe (Scotland) Ridge. It then fills much of the North Atlantic Basin, and rides up on the AABW on its way to the south, where it eventually joins the Circumpolar Current. During the last Ice Age the production of NADW probably ceased or was greatly reduced, for at least two reasons. One, influx of necessary saline waters from the North Atlantic was reduced, and two, the Norwegian Sea probably was crammed with pack ice which blanketed the water and prevented rapid cooling. Thus, the deep circulation in the Atlantic (and

104

probably also in the Pacific) was entirely different during glacials from what it is today.

On top of the cold, deep waters there is another layer before we reach the warm surface waters. This is the *intermediate water,* which sinks in the subarctic and subantarctic open ocean convergences and forms the bottom for the subtropical gyres. Along the ocean margins, where the intermediate waters intersect the continental slope, the flow tends to be in a direction opposite to that of the surface waters.

4.3.5 Exchange Currents. Geologically highly significant are the currents which provide for the exchange of waters between marginal, semi-enclosed seas, and the open ocean. The exchange entirely dominates the chemistry and fertility of the marginal basin and hence its sedimentation (see Sect. 7.6).

In arid zones, excess of evaporation over precipitation produces heavy water in the marginal sea; as it flows out over the sill, it is replaced by surface water from the open ocean. This is the case in the Mediterranean, and this circulation is called *anti-estuarine.* The salty Mediterranean waters are found at 1500 m depth over the entire central Atlantic, and affect profoundly the development of abyssal and deep water masses. A similar situation obtains, on a smaller scale, for the Red Sea and the Persian Gulf. Here the saline waters sink to intermediate depths below the thermocline, and affect the development of the oxygen minimum in the Arabian Sea. Basins with *anti-estuarine circulation* tend to collect *carbonate* on the sea floor, and tend to discriminate against organic carbon, opaline silica, and phosphatic deposits.

The influx of Atlantic surface waters into the Mediterranean is a strong current and a huge river: 50 times the Mississippi at flood stage. It is said that the Battle of Trafalgar (1805) owed its outcome to the Gibraltar Current. The reason given is that the French and Spanish fleet had to wait for favorable winds to buck the Current, while Nelson was able to assemble his fleet and poise it for strike. In modern times the submarines invading the Mediterranean had to struggle against the deep outflow. Already in 1820, the British Admiral W. H. Smyth observed the in- and outflow through the Gibraltar Straits, and gave the correct explanation. It was not, however, generally accepted for another 50 years of discussion.

The inverse situation — shallow current out, deep current into the basin — is typical for estuaries and fjords, and occurs on a large scale in the Black Sea and the Baltic Sea. Excess precipitation over evaporation is necessary to develop this circulation. At the Bosporus, the connection between the Black Sea and the Mediterranean, fishermen have known since time immemorial that their nets would pull toward the Mediterranean at the surface and toward the Black Sea at the bottom. L. F. Marsili (1681), pioneer in marine geology, confirmed the effect by measurement and correctly explained the currents as being due to density differences. The light excess water flows out of the Black Sea at the top; the heavy saline water from the Mediterranean falls toward the bottom of the Black Sea, displacing the less saline water above. Ultimately, anaerobic conditions result from this circulation, called *estuarine,* and organic-rich deposits accumulate on the sea floor. The hows and whys are discussed in Section 7.6.

4.3.6 Turbidity Currents. We have already mentioned mud-laden gravity-driven currents in connection with the origin of continental slopes, submarine canyons, and deep sea fans (Sect. 2.11). The realization that these currents play an extremely important role in present-day sedimentation as well as in the geologic record, came only in the 1950's, largely through the work of Ph. H. Kuenen (Fig. 4.24a).

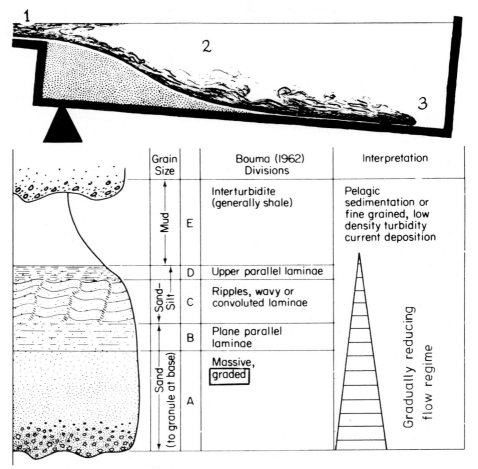

Fig. 4.24a, b. Origin of graded layers.

a Experiment of Ph. H. Kuenen. *1* Turbid, sediment-laden water is introduced into the tank; *2* water in tank remains still and clear, except over the bottom, where the denser muddy water rushes downslope; *3* turbulent front of the turbidity current. Depending on its strength a turbidity current can erode and redeposit enormous amounts of sediment. (From J. Gilluly et al., 1968, Principles of geology. W. H. Freeman, San Francisco, after photos by H. S. Bell, Cal. Tech.)

b Standard sequence of divisions in a turbidite layer, as proposed by A. H. Bouma. The *lower part* is the graded bed, produced by a turbidity current. The *upper part* results from "normal" sedimentation; it contains almost all the geologic time represented. (After G. V. Middleton, M. A. Hampton, 1976, in D. J. Stanley, D. J. P. Swift, Marine sediment transport and environmental management. John Wiley, New York)

106

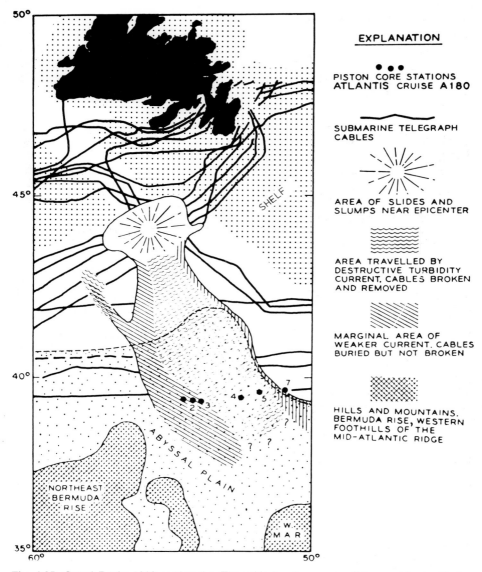

Fig. 4.25. Grand Banks 1929 earthquake. The cable break sequence (later combined with the stratigraphic record in the cores) was interpreted by B. C. Heezen and M. Ewing (1952, Am J Sci 250: 849) as evidence for high velocity turbidity currents. (From B. C. Heezen, in M. N. Hill, 1963, The Sea 3: 744. The land area, in black, is Newfoundland)

Kuenen established, by experiment, that such heavy downhill flowing currents can exist in nature and that they would deposit layers which show the *grading* familiar from hitherto unexplained sediment-series *(flysch)* in the Alps and other mountain ranges (Fig. 2.24b).

107

Before the *turbidite hypothesis* became a standard tool in the interpretation of sediments, the "desolate thousand-fold alternation of sandstone, limestone, and shale" which faced the alpine geologist in the flysch was a complete mystery. Now these alternations are generally regarded as a record of pelagic sedimentation with periodic influx of mud-laden bottom-hugging water bodies. These muddy waters, upon decelerating, drop their load within a short time as *graded layers*. Such layers are found within the fan deposits of continental slopes (Sect. 2.11), and in the sediments building the abyssal plains (Sect. 8.5).

Kuenen's work inspired marine geologists to look for direct evidence of the action of turbidity currents in the present ocean. A well-known example of this search is the investigation of the succession of telegraph cable breaks down the continental slope, which occurred after the 1929 Grand Banks Earthquake off Newfoundland. From the timing of the breaks (as recorded by the telegraph companies involved) B. C. Heezen and M. Ewing concluded (1952) that the quake had set off turbidity currents which moved downhill at high speed and snapped the cables (Fig. 4.25). The speeds they estimated were on the order of 25 to 50 miles per hour (10 to 20 m/s) — velocities of powerful super-currents. (For comparison, the fastest normal ocean currents run at about 2 m/s, or 5 miles per hour). While their calculations are not generally accepted, their studies did raise the possibility that such currents exist.

If currents like this occurred throughout geologic time, the origin of submarine canyons, of submarine fans and their valleys, and of the thick deposits below abyssal plains becomes somewhat less of a mystery. The possible velocities and densities of marine turbidity currents (as yet not observed directly) are still a matter of discussion.

5 Sea Level Processes and Effects of Sea Level Change

5.1 Importance of Sea Level Positions

When studying sedimentary rocks on land, the first question a geologist will ask is whether the sediment was laid down above or below sea level, that is, whether or not it is of marine origin. For marine sediments, the next question usually is about the depth of deposition, that is, about the *position of sea level* relative to the sedimentary environment. On the present sea floor, depth of deposition rather dominates the major facies patterns of the material accumulating on it: the size distributions of clastic sediments, the chemistry of biogenous and authigenic matter, the distribution of benthic organisms. For the past, *sea level fluctuations,* on scales between thousands and millions of years, dominate the calender of geologic history (Fig. 5.1).

Sea level fluctuations are of two kinds: *global* and *regional.* Global fluctuations produce contemporaneous transgressions and regressions on the shelves of all continents. These changes in sea level are called *eustatic,* they originate from changes in the volume of ocean water or in the average depth of the ocean basin. Examples are changes in ice volume, or changes in sea floor spreading rates, as we shall see. Regional fluctuations consist in transgressions and regressions on one particular shelf; they are produced by regional sinking or uplift of the shelf. Hence, such sea level fluctuations are called *tectonic,* with the understanding that the tectonics are of regional importance only.

Where the sea level intersects the continental margin, physical, chemical, and biological processes are of high intensity. Waves, tides, and currents show maximum activity. The productivity of tidal lands and of the littoral is exceptionally great, and the sediment is intimately associated with rapid nutrient cycling, gas exchange, and life processes in general. Furthermore, sea level is the baseline of erosion and deposition: exposed areas erode, submerged areas build up. The erosional and depositional processes at and near sea level to a large extent determine the *coastal morphology* we see. They also leave their distinct imprint in the record; they are *sea level indicators.*

On a larger scale, the position of sea level with respect to the global hypsographic curve is of great importance. It determines the degree to which shelves are submerged. Submerged shelves absorb much more sunlight than exposed ones. In a very simplified way, one might say that climate is mild during times of high sea level, harsh during times of retreat of the ocean into its basins. Thus, sea level fluctuations are closely tied to *paleoclimatic evolution.*

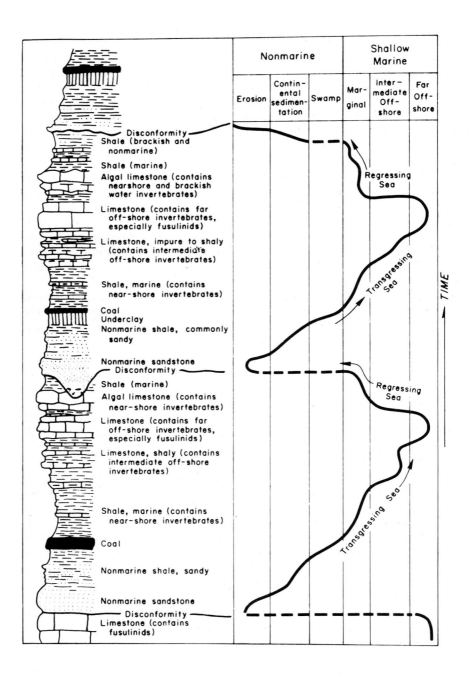

Fig. 5.1. Sea level fluctuations: calendar of geologic history. The diagram shows a section of the late Paleozoic cyclothem successions in Kansas, which consist of alternating marine and nonmarine sediments. Description of facies to the *left;* interpretation in terms of depositional environment to the *right.* (After R. C. Moore as drawn by J. C. Crowell, 1978, Am J Sci 278: 1345)

Quite generally, the rates of erosion of the continents and the sites of deposition in the ocean depend much on continental relief and position of sea level with regard to the shelves. How sediment is transported to the deep ocean also is controlled by sea level. During high stands, the transport of sediment by turbidity currents will be reduced. Turbidity currents depend on supply of mud to the outer edge of the shelf. This supply is highest when sea level is low, and is greatly reduced during high stands when the submerged shelves trap the material delivered by rivers. Thus, the types of *sediment bodies* found at the base of the *ocean margins* should depend on the history of sea level fluctuation.

Changes of sea-level positions through time, then, are of fundamental importance in the geological sciences. The dynamics of coastal morphology, the mapping of sea-level indicators, the interaction of sea level and climate, and the origin of sediment bodies in continental margins are topics which must be considered. However, sea level also is of the utmost interest to people living at the coast: their livelihood and their very survival depend on various manifestations of sea level change. This is true for both short-term fluctuation (tides, storm waves, tsunamis) and long-term trends.

We shall next review the evidence for sea level processes and fluctuations, starting with the short-scale and the obvious, and proceeding to the Pleistocene and pre-Pleistocene changes from there. At the end we append a story on Venice: the old city is slowly but surely sinking below the sea.

5.2 Sea Level Processes and Indicators

5.2.1 Wave Action. The most conspicuous indicators of sea level are those associated with wave action. Familiar examples are wave-cut terraces, and all kinds of beach deposits. We have already discussed some of the pertinent processes (Sects. 4.2.1 and 4.2.2).

Wave action leaves both erosional and depositional witnesses in the record. Many coastlines are marked by a steep escarpment where the land ends and the sea begins. Such escarpments, or *sea cliffs* are the result of wave attack. The wave-cut platform in front of a cliff is the most conspicuous lower boundary of wave-erosion. It may be thought of as the floor of a series of notches and caves which are cut into the retreating cliff (Fig. 5.2). The rate of growth of such a platform varies greatly; it depends on the force of the waves, the resistance of the cliff material, and the time available for cutting. In Southern California, rates of cliff retreat are between 1 and 100 feet per century, for example. Rates of sea cliff retreat on the order of 1 m/yr are common on certain coasts of the North Sea in the Atlantic, where stormy winter seas eat their way into unconsolidated glacial deposits. In England, many coastal villages have been lost through the centuries, due to erosion.

Waves transport and rework the sediment, changing its texture by sorting, and influencing its structure, for example by producing ripple marks (Fig. 5.3). Caution is indicated, however, since ripple marks also occur on the deep sea floor, far below sea level (Fig. 5.3b). Thick enrichments of heavy minerals, so-called *placers,* depend on a process which only exists on a beach (Sect. 10.3.3).

111

Fig. 5.2. Wave-cut terrace at Enoshima, Pacific coast of central Japan. (Photo E. S.)

Well-sorted skeletal remains — shell pavements or *coquina* — also are indicators of wave action. Waves usually influence the sea floor only down to about 10 to 20 m, and even great storm waves have a wave base no deeper than about 30 m.

Waves can also be responsible for *producing* certain types of sediments, mechanically as well as chemically. Polished beach shingles and calcareous oolites are examples. Calcareous oolites are characteristic for unusually warm and saline environments. Apparently they grow only within the zone agitated by waves and tides.

5.2.2 Tides and Storm Action: the Intertidal Zone. The tides, those long waves made by the Moon's daily passage, raise and lower the sea level along the coasts by a few centimeters or by many meters, every day. The tides originate in a complicated interplay between the rotation of the Earth–Moon system about its common center, the system's rotation about the sun, the Earth's spin, and the topography of the oceans and seas. The tides which we witness at the coast are large rotational waves radiating from points of no motion situated in the central areas of the large ocean basins. The amplitudes of the waves are highest close to the coast (Fig. 5.4).

Thus, tides are strictly tied to astronomical forcing and are shaped by basin morphology. If the distribution of their frequencies and amplitudes can be reconstructed for the geologic past, we can obtain important clues about changes in the Earth–Moon system, and about basinal morphology. Some work has been done on the shell-structure of tidal organisms, to learn to "read" shells in terms of tidal records, but hardly enough for quantitative geologic application.

Here we are concerned with the "indicators" of tidal action in a more general sense. Surely the classic sea level indicator facies is represented by the whole of

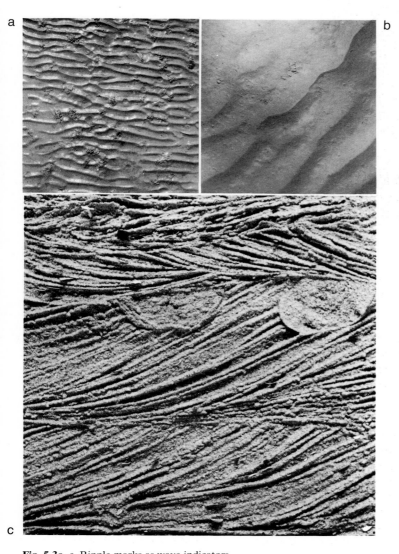

Fig. 5.3a–c. Ripple marks as wave indicators.
a Oscillation ripples, wave length about 5 cm, with *Arenicola* fecal strings; tidal flats, North Sea. **b** Oscillation ripples, wave length 30 cm. Top of Sylvania Sea Mount near Bikini Atoll, depth 1500 m, calcareous ooze. **c** Internal structure of current ripples, sandy tidal flats, German North Sea Coast. Frame is 22 cm wide. (Photos E. S. **a**, H. W. Menard **b**, H. E. Reineck **c**)

113

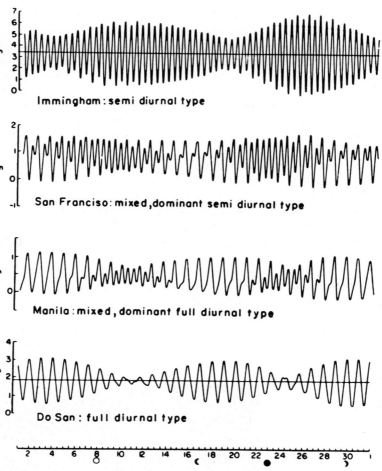

Fig. 5.4. Tidal records from four harbor cities. At *bottom* the dates (March 1936) and the phases of the moon. (From A. Defant, 1966, in G. Neumann, W. J. Pierson, 1966. Principles of physical oceanography. Prentice-Hall, New Jersey)

the deposits of the intertidal zone, which is alternately submerged and exposed by the tides. Sinking coasts with high sediment supply tend to have large intertidals, and hence accumulate large intertidal sediment bodies. We have already introduced one such type of broad intertidal, the "wadden" (Fig. 4.18; Sect. 4.3.2). Others will be discussed in Section 5.3. Intertidal flats and adjacent areas are subject to rapid changes in the conditions of sedimentation. Small sea level changes, or geographical changes in tidal access, can quickly lead to submergence or emergence. Storms play an important role in producing such change.

Peak winter storms wreak havoc along the coastal lowlands of the North Sea from time to time. In 1362, an enormous storm raised the sea level by about 6 m along the Frisian Coast, where Germany borders with Denmark. The storm

114

flood created new access for the tides into the salt marshes and moors lying behind natural barriers, inland from the intertidal flats. Soon the tides brought mud which accumulated on top of the marshes and the peat, producing a *storm layer*. To the people living there, this geologically common event meant destruction of homes and pastures. In 1634, another fierce storm produced widespread flooding, destroying villages and killing thousands. The present coastal geography of the area is essentially a product of those two storms. A faint indication of plowed fields can be seen to this day on certain intertidal flats of the region. Some of the land has been reclaimed since, aided by the influx of sediment from the wadden, and by the building of dikes.

In areas such as these lowlands, then, slight changes in sea level, and the action of heavy storms, bring marked changes in the type of sediment deposited. Typically, the record which is preserved on the slowly sinking floor consists of an intercalation of peat, salt marsh deposits, marine muds, and the sands and shells of the beach. Storm deposits are common within the intercalation. Similar sequences are quite familiar from the geologic record (see Fig. 5.1). They are characteristic sea level deposits in areas with high terrigenous sediment supply. The sea level controls the groundwater level by damming up and floating the freshwater. The ensuing lack of drainage can produce swamps from which peat bogs develop. In the geologic record these appear as coal beds, between marine sediments.

In muddy intertidal flats of the subtropics, high evaporation rates can produce desiccation cracks (Fig. 5.5). Other clues to an intertidal environment of

a   b

Fig. 5.5a, b. Desiccation cracks in algal mats, in a lagoon on the west coast of Baja California near San Quintin. **a** View from a sand dune, toward inner lagoon. Marsh vegetation, stressed by high salinity from evaporation, gives way to algal mats toward the intertidal. **b** Detail of algal mats (section 2 m wide). The mats separate into polygonal pieces and curl upon drying. Upturned rims collect evaporites. (Photos E. S.)

deposition are rain drop impressions, *pseudomorphs* of cubic halite crystals (that is, infilling of crystal forms by sand or mud), precipitation of gypsum, tracks of land animals on sediments with marine organisms. Any of these clues can be covered up and preserved by flood deposits from the hinterland, in areas where the tidal flat is part of a delta setting. The delta flat, in fact, is an excellent marker of sea level. Commonly it is characterized by distributary channels, cut-and-fill, and the intermingling of marsh-lagoon-beach deposits (Sect. 5.3.2).

5.2.3 Photosynthesis. A large number of benthic organisms can indicate shallow water by their presence, but none as convincingly as those dependent on light. Photosynthesis can only proceed when sufficient light is available. The light intensity drops to 1% of the surface value anywhere between 10 and 200 m water depth — depending on the clearness of the water. Sessile plants, such as the geologically important calcareous algae and algal mats, generally occur no deeper than 100 m or so. Animals living in symbiosis with algae also indicate shallow water. These animals include large foraminifera, stone corals, even certain molluscs (Sect. 6.1.2).

5.3 Coastal Morphology and the Recent Rise in Sea Level

5.3.1 General Effects of Recent Sea Level Rise. The coastal regions of uplifted margins are largely shaped by the interplay between tectonic forces and marine processes, especially wave erosion. Raised marine terraces are a characteristic morphologic feature of such uplifted margins; the West Coast delivers many fine examples. In contrast, the coastal morphology of slowly sinking margins is entirely dominated by the processes associated with the sea level itself. In North America, the Gulf of Mexico and the East Coast provide prime examples of this type of morphology.

In order to understand the coastal landscape, we must be constantly aware of one central fact: the recent rapid rise in sea level which began about 15,000 years ago and lasted till about 7000 years ago (Fig. 5.6). The sea level rose in response to the melting of glacial ice, chiefly the Laurentian and the Scandinavian Ice Sheets (Antarctic melting is thought to have contributed less than 10% of the rise). The maximum addition of water to the ocean occurred somewhere between 13,000 and 9,000 years ago. The sea level changed by approximately 130 m on the whole; that is, it rose from the average depth of the shelf edge to the present positions.

The effects of the deglacial transgression on coastal morphology and sedimentation were varied and profound. On gently sloping shores the sea advanced very rapidly. In the upper Persian Gulf, for example, the coastline must have retreated by some 100 m per year, during the maximum rise of sea level. Whoever inhabited the Shatt-al-Arab at the time must have been very aware of the invasion of the sea.

In temperate humid regions, coastal peat bogs grew upward with the rising groundwater level; finally they became flooded by salt water and covered by

116

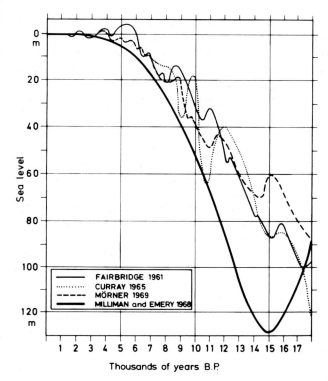

Fig. 5.6. Sea level rise during deglaciation. The diagram shows various hypotheses of how the sea level is supposed to have risen during deglaciation (about 15,000 to 9,000 years ago), and during the Holocene afterwards. The exact sequence and the periods of arrest (or perhaps regression) are not agreed upon, due to difficulties of dating (by ^{14}C), and regional tectonic overprint (rising or sinking shelfs and coasts). (From E. Seibold, 1974, in R. Brinkmann (ed). Lehrbuch der allgemeinen Geologie, vol/1, 2nd ed, F. Enke, Stuttgart)

marine sediments. Dunes were eroded away by the approaching surf, except where cemented by lime derived from shell material. Resistant matter left from the erosional process collected as a *transgression conglomerate* below the march of the waves, the *basal conglomerate* typical of many marine transgression sequences in the geologic record.

The exact time sequence of the rise of sea level during the melting of the northern glaciers is still a matter of controversy, as the various interpretations in Fig. 5.6 indicate. Was it an even, rapid change? Did it occur in pulses, as indicated by terraces on many shelves and coral islands? What is the local imprint of vertical coastal motion in each sea level curve? In northern latitudes, what exactly was the regional uplift of the crust in response to the removal of ice? How much was the crust depressed by the addition of water on the shelf? How did the Earth's rotation respond to the shift of weight from Canada and Scandinavia to the open ocean? How did this shift influence the *geoid*, the shape of the Earth and its sea level? These questions — and others — must be solved if the *sea level curves* are to be translated into *mass of water added per time*.

117

An outstanding example for the effects of the rise of sea level is the grand morphology of the East Coast of the U.S.A. It is a coast of drowned rivers (Fig. 5.7a). Longshore sand transport (Sect. 4) tends to close off the estuaries which fill with sediment to make marshlands (Fig. 5.7b).

Drowned rivers can make good harbors: the East Coast is dotted by ports with excellent deep-water access. Drowned rivers are sediment traps; they tend to

a

b

Fig. 5.7a, b. Sea level rise and East Coast morphology. **a** Drowned river valleys, Cape Cod. Barrier beaches migrate into the entrance of the estuary, as spits, and restrict access to the open sea. **b** Close-up of a migrating spit, Buzzard's Bay. Migration is to the *left*. Note forced detouring of the tidal channel. (Photos courtesy D. L. Eicher)

118

catch in their estuaries whatever material is brought from the continent. On the whole, therefore, the shelf off the East Coast is starved of sediment. What we find on the sea floor there is largely relict material from the last ice age. The distribution of such *relict sediments,* naturally, is not very indicative of the present-day processes active on the shelf, except through the evidence for reworking.

With this general background, let us now consider the geological-morphological processes at sea level in more detail, first at river mouths, then at low-land coasts between the rivers.

5.3.2 River Mouths. A glance at a world map shows that river mouths can be indentations or protrusions in the coastline — that is, estuaries or deltas (Fig.

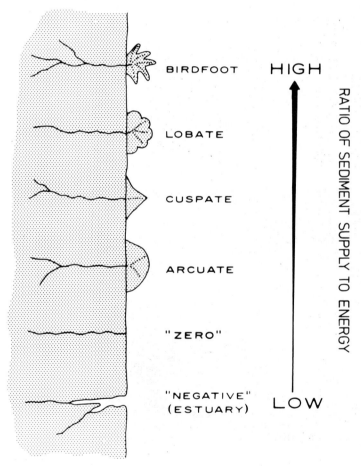

Fig. 5.8. Subaerial delta shapes as a function of sediment supply and distributional energy (waves, currents). (From J. R. Curray 1975, in A. G. Fischer, S. Judson (eds) Petroleum and global tectonics. Princeton Univ Press, New Jersey)

5.8). Tidal motion can reach far into the estuaries. Commonly, salt water intrudes along the bottom. This intrusion of the marine realm brings marine sediments upriver. Such sediments are easily recognized within the river deposits: echinoderm remains, foraminifera, marine ostracods.

Thus, both the river and the sea brings sediment. Why then are the estuaries not filled in? River floods, tidal action, and especially the young age of estuaries are the cause. The drowned river valleys produced by the transgression have not yet come to equilibrium with the sediment load. Where the river valley runs across the shelf and into a submarine canyon on the slope, the filling-in of the river mouth is slowed, because sediment can travel out of the estuary onto the deep sea floor. The Congo and the Hudson are examples. The best harbors, therefore, are those with a canyon offshore like New York, and those where strong tidal action keeps the outer river channel open, as in London, Bordeaux, or Hamburg.

For the geologist, who must learn to read the record, the river mouth of deposition, the delta, is of special interest (Fig. 5.9 and 5.10). Why a delta forms and not an estuary depends on many factors: for instance, low tidal activity as in land-surrounded seas (Mississippi and Nile deltas); high sediment load due to strong seasonal rains, and high erosion rates in the mountainous hinterland (the Indus, Ganges, and Irrawadi deltas).

When coast and sea level are stable, the delta builds out to sea, the marine facies retreats, the record shows *regression*. A borehole in such a delta shows

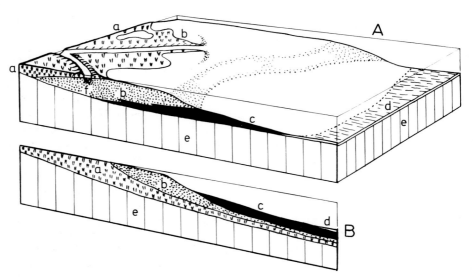

Fig. 5.9. Schematic cross-section through a birdfoot delta.
a Inter-levee lowlands, marshes (levees are ridges confining the distributary channels); **b** delta front; **c** prodelta; **d** open shelf floor; **e** older base (may or may not be deltaic); **f** distributary channel with fill and levees. *A* delta builds up and out, the sea retreats, the sedimentary sequence is "regressive" (a across b, b across c, etc). *B* The sea level rises and the delta retreats, the sedimentary sequence is "transgressive" (a below b, b below c, etc.). Note that the regressive or transgressive nature of the sequence cannot be seen on the surface, but only by studying the sequence of sediments within the delta

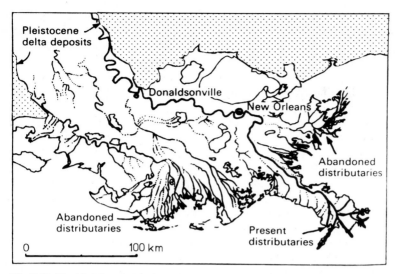

Fig. 5.10. The bird-foot delta of the Mississippi. The subsidence of the delta allows the sea to re-invade the areas of abandoned distributaries. (From J. Gilluly et al., 1968, Principles of geology. W. H. Freeman, San Francisco, after H. N. Fisk)

a *regressive sequence:* shallow deposits over deep ones (Fig. 5.9A). If sea level falls, of course, the regression is accelerated. Conversely, a rise of sea level or a sinking coastline will ideally produce a *transgressive sequence,* which is the reverse order of the regressive one. The recent rise of sea level resulted in the predominance of transgressive sequences on the top of the deltas of the world.

5.3.3 Lagoons and Barriers. Next to deltas we usually find low-lying coasts showing a facies zonation parallel to the coastline. Offshore bars, or barrier islands, commonly occur together with a sand beach (Fig. 5.11). The beach may be backed by dunes, which the wind piled up using sand from the beach. In turn, this beach–dune complex may form a barrier for interior lagoons, as is the case along much of the Gulf of Mexico and the East Coast. The narrow barriers can be tens of miles long. Rivers emptying into the lagoons may cut one or several channels through such barriers, especially when flooding, chopping them up into a series of barrier islands. From the seaward side, storm waves may break through the barrier, forming *overwash fans* on the lagoon beach (Fig. 5.12). Tidal action keeps such channels open, building deltas on both sides.

How does barrier-and-lagoon morphology reflect changes in sea level? Again, the balance between the rise of sea level relative to the land and the supply of sediment to the coast is the crucial factor to consider. Since the sea level stabilized, about 6000 years ago, supply of material has been all-important. Off Galveston Bay, for example, the high supply of sand allowed a rapid building-out of the offshore barrier (Fig. 5.13). On the whole, however, barriers and lagoons must have migrated landward over the last 15,000 years.

121

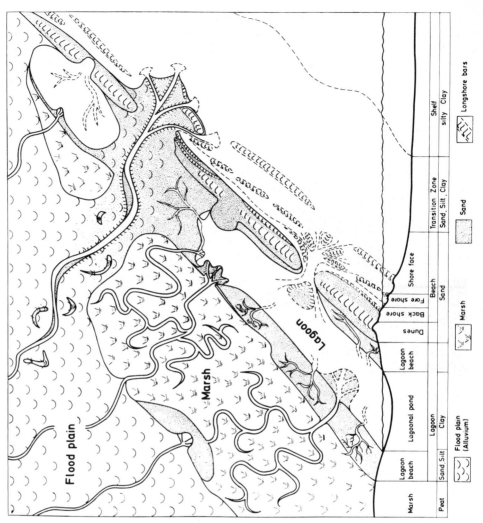

Fig. 5.11. Typical geomorphic features of a barrier-type coast. (From H. E. Reineck, I. B. Singh, 1973, Depositional sedimentary environments. Springer, Heidelberg, based on a diagram by C. D. Masters, 1965)

The barrier-type coast is very abundant: of 244,000 km coastline its share is about 32,000 km, that is 13%. North America and Africa each have about 18%, Europe only about 5%. A stable broad shelf and a high supply of sediment would appear to be favorable for the development of barrier coasts, while coastal uplift, narrow, dissected, and starved shelves seem unfavorable.

5.3.4 Mangrove Swamps. During the recent rise of sea level, generally speaking, the various facies zones paralleling the coast migrated landward. In the tropics,

one of the most impressive migrations of this type must have been the retreat —
and landward invasion — of the mangrove swamps. Mangrove growth dominates
many intertidal zones in the tropics (Fig. 5.14). Mangroves need year-round
temperatures of 20° C or higher. In equatorial regions with high annual rainfall,
for example off Cameroon and Guyana, mangrove forests form a broad belt

Fig. 5.12. Overwash fans on St. Joseph's Island, Texas, produced by storm waves. Open Gulf to the
left, lagoon to the right. (Photo courtesy D. L. Eicher)

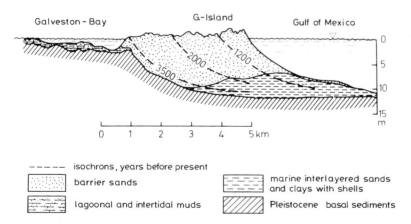

- - - - - isochrons, years before present

[barrier sands]

[lagoonal and intertidal muds]

[marine interlayered sands
and clays with shells]

[Pleistocene basal sediments]

Fig. 5.13. Architecture of Galveston Island. Regressive sequence below the barrier beach near
Galveston, Gulf of Mexico. Advance of barrier island is due to high supply of sand and relatively stable
sea level. The dates (1200, 2000, 3500), are based on the ^{14}C content of shells. (After J. R. Curray,
1969, in D. J. Stanley (ed) The new concepts of continental margin sedimentation. Am Geol Inst,
Washington DC)

123

a ... b

Fig. 5.14a, b. Mangrove vegetation on Bimini (Bahamas).
a *Rhizopora* at low tide. The roots are covered at high tide. Plant about 2 m wide. **b** *Avicennia* with air roots. Width of view at proximal edge 2.5 m. (From E. Seibold, 1964, *Neues Jahrb Geol Palaeontol* Abh 120: 233)

offshore. Here the mangrove forests blend into the tropical rain forests at high tide level. The expansion (and successive burial) of mangrove swamps during deglaciation due to a rising sea level, must have produced an organic-rich layer on many tropical shelves, much as peat growth did on temperate shelves. Such layers, when buried, will coalify. The coaly layers in ancient cyclothems (Fig. 5.1) apparently originated in similar fashion. Rapid burial of carbon extracts CO_2 from the atmosphere. The question of possible CO_2-changes due to changes in sea level is a topic of active research in marine chemistry and geology (see also Sect. 8.3.6).

5.4 Ice-Driven Sea Level Fluctuations

5.4.1 The Würm Low Stand. In the last section we referred to the rapid rise of sea level, between 15,000 and 7000 years ago, which was caused by the melting of glacial ice (see Fig. 5.6). This rise is but one phase in the long series of sea level fluctuations of the Pleistocene. Sea level has been constantly changing over the last several hundred thousand years as a result of the waxing and waning of large continental ice masses. The rise of the last deglaciation, which was witnessed by our ancestors, was one of the biggest and fastest sea level changes ever, as far as we know.

It was the result of a maximum change of climate: from a peak cold period to a peak warm period.

About 17,000 years ago, during the last major ice age (the *Würm* in Europe, the *Wisconsin* in North America), enough water was tied up in the continental glaciers to depress sea level by some 130 m. Large shelf areas fell dry as a result of the ice-caused regression of the sea. Rivers crossed the shelves and entered the sea at the shelf edge, cutting backward into the shelf. Their sediment load was dropped in a narrow zone at the upper slope, became unstable there and slid,

124

starting turbidity flows which rushed down the undersea canyons. In front of the ice, outwash plains (or moraine ridges next to the terminal ice tongues) developed on the exposed shelf. Dune fields evolved on shelves where the climate was favorable.

Land animals spread over the emerged shelves and used them as land bridges in some cases, for example, between Northeast Siberia and Alaska. Mammoths roamed over the large area covered by the North Sea today. Their remains are found on the sea bottom now, together with the tools of prehistoric hunters. Could these people of the late Stone Age have left a memory of the effects of rapid sea level rise, in the ubiquitous legends of a Great Flood?

5.4.2 Pleistocene Fluctuations. There is an indication that, throughout the Pleistocene, major transgressions were much more rapid than any of the regressions. In general, apparently, the build-up of ice (and hence the sea level drop) proceeded at a slower rate than the melting of ice. The evidence for this concept rests on the oxygen isotope record of pelagic foraminifera (Fig. 5.15).

How can we measure sea level changes in the chemistry of foraminiferal shells? The principle of this method was introduced by C. Emiliani, in 1955.

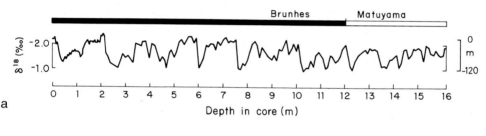

a

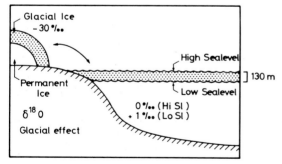

b

Fig. 5.15a, b. Fluctuation of $\delta^{18}O$ in the planktonic foraminifer *Globigerinoides sacculifer,* western equatorial Pacific. **a** "Brunhes": present normal magnetic epoch, since 700,000 years ago. "Matuyama": previous epoch, during which the Earth's magnetic field was reversed. Seven isotope cycles are visible within the Brunhes, the average duration is 100,000 years therefore. The temperature of surface waters is nearly constant through time, in the area where the core was taken. Hence, the isotope fluctuations are produced largely by the build-up and decay of northern ice-masses, as shown in **b.** (Core data from N. J. Shackleton, N. D. Opdyke, 1973, Quat Res 3: 39)

125

The shells of the foraminifera consists of calcium carbonate, $CaCO_3$. Hence, they contain oxygen. The water within which the shells grow, H_2O, also contains oxygen. There are three kinds of oxygen, the normal one, with atomic weight 16, and two rare ones, with atomic weights 17 and 18. The oxygen-17 is very rare and we need not pay further attention to it. Through mass spectrometry, one can determine the ratio between oxygen-16 and oxygen-18 (commonly written ^{16}O and ^{18}O). The ratio of these two isotopes in the shells is in equilibrium with the ratio of the isotopes in the water in which the shells grew. In other words, if the $^{18}O/^{16}O$ ratio in the water changes, it will change similarly in the shell.

This fact relates to sea level change as follows. Every time the sea level drops, the $^{18}O/^{16}O$ ratio of the water increases, because the glacial ice is made of water which is impoverished in the isotope ^{18}O. Hence, seawater is enriched in ^{18}O during

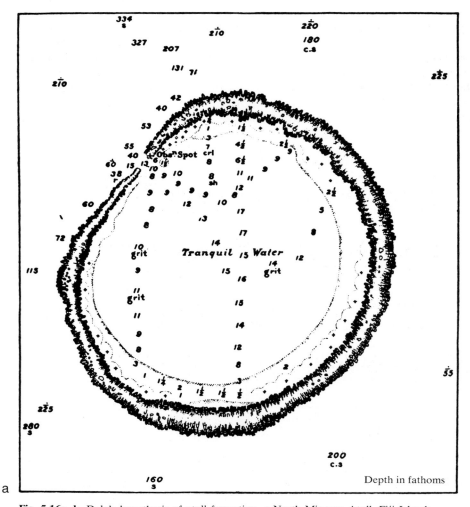

a

Fig. 5.16a, b. Daly's hypothesis of atoll formation. **a** North Minerva Atoll, Fiji Islands.

126

glacials (Fig. 5.15b). The $^{18}O/^{16}O$ ratio of the carbonate shells reflects this change in the chemistry of seawater. Temperature also influences the $^{18}O/^{16}O$ ratio of the shells. In Fig. 5.15a, however, the long sediment core on which the measurements were made comes from an area where the effect of temperature on the isotopic signal can be neglected in the present context.

Note that the oxygen isotope curve (Fig. 5.15a) indicates that sea level fluctuated between rather well-defined limits. Sea level was never much higher than right now, and it was never much lower than during the last glacial. There must be a climatic factor or factors which prevent a build-up of ice beyond a certain limit, and other factors which prevent additional melting once a certain amount of ice has been melted. We shall return to this puzzling observation in Chapter 9.

5.4.3 Effects on Reef Growth. The sea level fluctuations of the Pleistocene left their imprint also in the shallow water carbonates, notably in the tropical reefs. Each highstand of the sea level resulted in a build-up of reef carbonates, while the lowstands resulted in erosion. In fact, the origin of atolls, those ring-shaped islands dotting the central Pacific, has been contemplated under this aspect (Fig. 5.16). Thus, while Darwin's hypothesis of submergence is correct in a general sense (see Sect. 7.4.3, Fig. 7.9), the influence of the fluctuating sea level must not be forgotten.

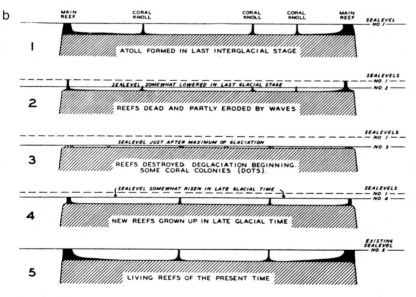

b Daly's Glacial Control Theory of coral reefs. The time sequence is from *1* (124,000 years ago) to *5* (the present). The ring shape of atolls is due to the more favorable situation for coral growth at the edges of the island (due to cleanness of water, and high food supply). The knolls in the lagoon grow up on slightly elevated, mud-free ground (From R. A. Daly, 1934, The changing world of the ice age. Yale Univ Press, New Haven)

On rising shores with reef belts, such as Barbados in the Caribbean, the fluctuating sea level translates into raised reef terraces. The terraces correspond to the highstands in sea level. They have been dated, by measuring the concentration of radioactive uranium within individual coral heads, and the product of uranium decay, the element thorium. Since at the time of growth of the coral there is essentially no thorium present, one can tell how much decay of uranium took place from the ratio of these radioactive isotopes. Results show that the uplifted corals grew during several high sea level stands, namely 124,000 years ago, at 103,000 years, and at 82,000 years — very much as expected from the sea level curve shown in Fig. 5.15.

Ice-driven sea level fluctuation seem to have been less pronounced in the earlier Pleistocene than in the later part of the period. However, even back to the Miocene there were such fluctuations, since ice build-up apparently started on Antarctica some 15 million years ago. As far as we can tell from the oxygen isotope record (see Chap. 9), the strength of the sea level fluctuations increased at 6 million years ago (in the latest Miocene) and again about 3 million years ago, when northern glaciations set in.

5.5 Tectonically Driven Sea Level Fluctuations

5.5.1 Sea Level and Sediment Bodies. Sea level fluctuated considerably all through the Phanerozoic, even during periods when apparently no ice was present. As to the present sea floor, fluctuations since the Jurassic are of special interest: essentially they determined the sequence of sedimentary layers within the continental margins.

The thick sediment stacks in the "passive" continental margins have been much studied for economic reasons. When a margin sinks more or less continuously, coastal sediment bodies must reach great thickness, provided the sediment supply keeps pace with subsidence, and deposition remains locked in to the sea level. Tertiary sandy beach deposits can attain up to 1,500 m thickness in the northwestern Gulf of Mexico, for example, also widths of up to 40 km and lengths of several 100 km.

The significance of the sand bodies in economic geology lies in their porosity and permeability. Thus, they can retain (and deliver) great quantities of water, petroleum, or gas. It is for this reason that the origin, dimensions, and properties of coastal sand bodies have received much attention by marine geologists as well as oil geologists. Both drilling and seismic exploration helps define their extent in the coastal areas of interest.

To interpret the sequences of marine sediments on land, in the margin, and on the deep sea floor, and for economic reasons as well, we would like to know how sea level fluctuated over the last 150 million years. However, in as much as the sea level variations within this geologic period were not driven by the growth and decay of ice caps, they were not reflected in the isotopic composition of seawater. We cannot, therefore, find them in the isotopic composition of the foraminifera in the manner indicated earlier.

How then can we measure these fluctuations?

5.5.2 Reconstruction of Sea Level Changes. The intensive world-wide exploration of continental margins by seismic profiling, has recently led to the realization that the sediment-stacking patterns in margins of different ocean basins are quite similar — hence, it is assumed, they must be due to global sea level variation. Using this hypothesis, P. Vail, R. M. Mitchum, and their associates have developed a method to derive sea level fluctuations from the geometry of sediment layers, as recognized on seismic reflection records. The basic idea is this: during a relative rise of sea level (transgression), sediment layers expand into shallower water, and they become wider as they build up. During a fall of sea level (regression) the reverse occurs, and erosion sets in on the shelf. Erosion, of course, produces a *hiatus:* a surface which joins older and younger sediments in a discontinuous way. Hence, the course of regression is poorly documented, and it looks as though it happened rapidly. It is much like cutting a section out of a movie-film: the change between "before" and "after" becomes very sudden.

The depositional patterns during a highstand and a lowstand of sea level are illustrated in Fig. 5.17. In essence, the sediment pile moves into deeper waters during the lowstand, and its geometry changes correspondingly. The construction of the apparent sea level cycles is based on this changing geometry of sediment stacks (see Fig. 5.18). The falls of sea level appear as instantaneous events, due to the presence of erosional (or nondepositional) hiatuses. While sea level might recede quickly, it cannot do so instantaneously, of course.

The "Vail sea level curve" is a useful tool for the correlation of seismic stratigraphies of continental margin sediments. To what extent this curve reflects the true global sea level variations is not known. One problem is that the rate of sediment supply (which is a factor more or less independent of sea level) must

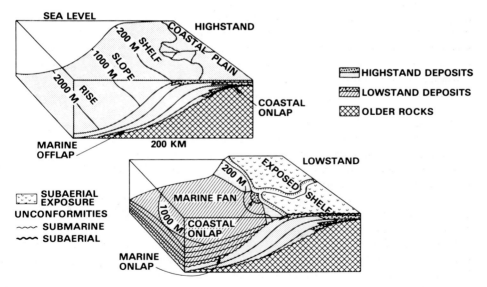

Fig. 5.17. Effect of sea level position on depositional pattern at the continental margin. Highstand, and lowstand of sea level. (From P. R. Vail et al., 1977, Am Assoc Pet Geol Mem 26: 49)

129

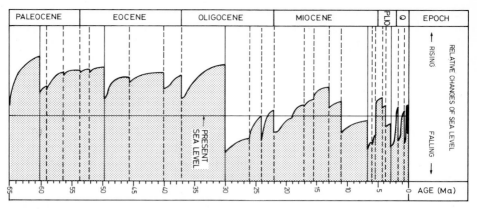

Fig. 5.18. Relative changes of sea level as deduced from the geometry of margin sediment bodies. The underlying model is that given in Fig. 5.17. (After P. R. Vail et al. 1977, Am. Assoc Pet Geol Mem 26: 49)

play an important role in controlling the geometry of the sediment bodies. Another is that the rate of sinking of the margins must be considered. Quite possibly, sea level fell continuously since the latest Cretaceous and only varied its *rate* of falling. When it fell slowly, the sinking passive margins overtook it, and accumulated transgressive sequences. When sea level fell quickly, the passive margins could not keep up: the result was regression.

5.5.4 The Causes of Change. If sea level fluctuates through time, we must search for the cause or causes of such fluctuations.

The obvious way to change sea level (if not by ice) is to change the average depth of the sea floor: if the floor shallows, the sea level rises; if it deepens, the level falls. We have seen that the depth of the sea floor is tied to its age: to change the depth, we must change the age. To decrease the age (and hence cause a transgression) we must replace old sea floor with young. This can be done by increasing the total mass of new lithosphere formed per year, that is, by increasing sea floor spreading rates or by increasing the length of the Mid-Ocean Ridge and the trenches, or both (Fig. 5.19).

Global tectonic events which led to a change in the average age of the sea floor, and hence its depth, apparently happened in the past. In the opening of the Atlantic, for example, young sea floor was generated by the new spreading center in the Atlantic, while old floor was subducted elsewhere in the Pacific (or else the entire globe would expand). As the Atlantic grows, the average age of its sea floor increases all the time — at some point it becomes older than the average age in the Indo-Pacific and hence starts increasing the global average. Further growth of the Atlantic then results in a drop of sea level. There are indications, from the magnetic stripes on the sea floor, that global spreading rates in the Late Cretaceous may have been much higher than now. It has been proposed that the high sea level stands of the Late Cretaceous were caused by such a fast spreading rate.

130

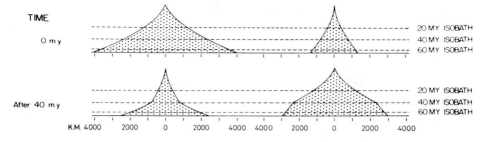

Fig. 5.19. Relationship between volume of Mid-Ocean Ridge and spreading rate. *Upper* cross-sections of ridges spreading at 6 cm/yr *(left)* and at 2 cm/yr *(right)*. *Lower* cross-sections of the same ridges 40 million years after a change from 6 cm/yr to 2 cm/yr (left) and a change from 2 cm/yr to 6 cm/yr *(right)*. The changes in ridge volume must produce corresponding changes in sea level. (From W. C. Pitman, 1979, Am Assoc Pet Geol Mem 29: 453)

The changes which can be produced by replacing old sea floor with young, and vice versa, are large but gradual. How can sea level be changed rapidly?

Mountain-building with shallow ocean crust, or with continental crust, is one way. In this process, shallow crust (which is stacked up within mountain ranges) is removed and replaced by deeper sea floor which will cover itself with a thicker layer of water, drawing down the general sea level. If the Tibetan Plateau represents "doubled" continental crust, for example, the corresponding sea level fall is about 40 m.

The quickest way to change sea level is to fill or to empty an isolated ocean basin. We know (thanks to Leg 13 of the Deep Sea Drilling Project) that the Mediterranean dried up intermittently between 5 and 6 million years ago. The water had to go elsewhere: global sea level was raised by about 10 m whenever the Mediterranean dried up. Conversely, when ocean water rushed in to fill an empty basin, global sea level must have dropped by the same amount. Since salt deposits have been found, or are thought to be present, in other ocean basins also (North Atlantic, South Atlantic), geologically instantaneous transgressions and regressions may have been quite common after the break-up of Pangaea.

Even a "geologically instantaneous" transgression would not necessarily be spectacular to a human observer. A transgression which is geologically very fast (although not produced by the hypothetical isolated basin effect) is proceeding right now in Venice. To this phenomenon we turn next.

131

5.6 Sea Level and the Fate of Venice

5.6.1 Venice is Sinking. Ancient Venice, with its San Marcus Cathedral, palaces, and canals, is slowly sinking below sea level.

Can Venice be saved? Let us take a closer look at the formidable scale of problems associated with the task of keeping the sea out. The necessary information has been collected by the *Laboratorio per lo studio della dinamica delle grandi masse* of the *Consiglio Nazionale delle Ricerche*.

Venice is at sea level. The city lies at the rim of the Po delta, in a lagoon protected by a barrier island, the Lido. The Lido is breached by tidal inlets (Fig. 5.20). The signs of sinking of the city are everywhere: docks must be built up; entrances are bricked in; steps are under water; during high water the city is flooded. The extent of sinking was determined by careful geodetic measurements starting from Treviso as a fixed point. The rate increased dramatically during the sixties. Between 1908 and 1925, Mestre-Marghera sank 0.15 mm per year, on the average. Between 1925 and 1952 it was 0.7 mm, and between 1952 and 1968, 3.8 mm per year. In parts of the region around Venice, the ground sank by 120 mm between 1952 and 1968, that is, 7.5 mm per year!

5.6.2 What Are the Causes for the Subsidence?

1. One reason is the general sinking of the Po Delta. By drilling, it was found that there are more than 3 km of Quaternary sediment below the central delta. If we set the Quaternary equal to 2 million years (see App. A3), we get an average rate of subsidence of 1.5 mm per year. Venice lies at the rim of the delta — hence it sinks a little more slowly, say, 0.5 mm per year on the average.

2. The Quaternary sediments underlying the region are partly marine and partly continental, indicating fluctuations in the position of the coastline relative to sea level. Tectonic movements and changes in sediment supply of the Po brought about these fluctuations. At the present time, the shoreline tends to advance inland.

3. Compaction of the sediments adds a component to subsidence, which can be locally reinforced by building and by infilling of lagoon areas. This may be the

132

reason for increased sinking around the city's railway station and harbor (Fig. 5.20).

4. The sand layers of which most of the underlying sediment consists, are groundwater reservoirs *(aquifers)*. When groundwater is removed faster than it is replaced, the ground sags. The sag can reach over considerable distance — a kilometer or more — if the pumping is done from depths of several hundred meters. In the last 20 years the groundwater level was lowered by 5 m in and around Venice, by 20 m in the industrial zone of Marghera. During this period, subsidence accelerated markedly. Perhaps this factor alone can account for the high sinking rates.

5. For the last 100 years the sea level has been rising globally by 1 to 2 mm per year, due to input from the melting of continental ice, presumably mainly in the Antarctic.

6. The normal tidal range is about 1 m at Venice. Storms can increase a high tide level considerably. On November 4, 1966, a storm tide reached +1.9 m, flooding the San Marcus Square (Fig. 5.21). Normally the lagoon would act as a buffer. However, large parts of the lagoon were filled in to gain land for industrial purposes, and shipping channels were deepened. The concern that these actions will increase the influence of storms and storm tides appears justified.

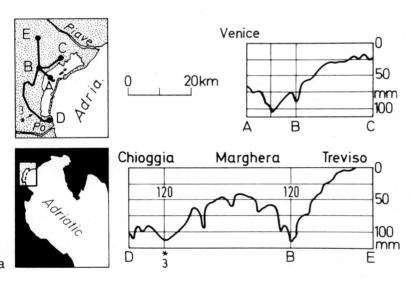

a

Fig. 5.20a. Amount of subsidence of ground in the vicinity of Venice. *Upper left* location map of Venice and surroundings. *Right* subsidence profiles along the lines *A B C (upper)* and *D B E (lower)*. Difference in elevation (in mm) refers to period between 1952 and 1968. Treviso *(E)* is taken as fixed point for the geodetic measurements; hence its change in elevation (if any) is defined as zero. Note high rates of subsidence (about 7.5 mm/yr) in the industrial region of Marghera (near *B*) and in the Po delta near Chioggia *(D)*. The center of the city of Venice subsided by about 80 mm, i. e., 5 mm/yr. (Data from R. Frasetto, 1972, CNR-Lab Stud Din Masse Tech Rep no 4). The postal stamps (opposite page) reflect the international concern about the fate of Venice. They picture the Palace at the Canale Grande and San Marcus Cathedral, threatened by the sea

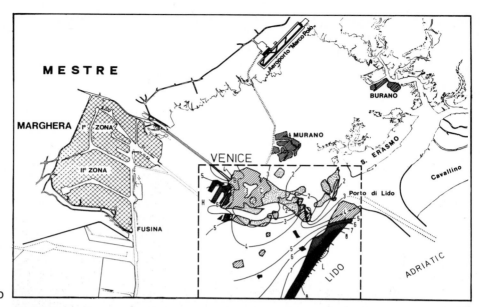

Fig. 5.20b. Lagoon of Venice. *Center* subsidence of Venice between 1961 and 1969, shown as contours in mm. *Stippled* land; *white* lagoon and Adriatic. Maximum sinking occurred on offshore barrier Lido (at *L*), due to building load and groundwater extraction. Also near the railway station *(B)* and in the harbor *(H)*, where filling and building took place. In the Canale Grande area *(C)*, local subsidence of building may be favored by erosion of the base through water motion from the traffic. (Modified after M. Caputo et al., 1972, CNR-Lab Stud Din Gr Masse Tech Rep no 9)

5.6.3 What Can Be Done About It? The results obtained from the studies summarized suggest that the pumping of groundwater must be regulated. Also, the effects of filling in of the lagoon, of deepening the channels, and of changing the shapes of the inlets (providing ready access for storm tides) must be studied. Of course, the problem of removing waste in an economic and sanitary fashion must not be forgotten: the tidal action flushes the city's polluted canals. The possibility of regulating tidal flow through sluices at the Lido is being investigated.

Nothing can be done about the general regional sinking, or the global rise of sea level. We cannot control such geologic factors; we must live with them.

134

Fig. 5.21. The flooding of San Marcus Square in November 1966. (Photo E. S.)

6 Organisms on and in the Sea Floor – Distribution, Activity and Environmental Reconstuction

6.1 The Ocean Habitat

6.1.1 Diversity of Organisms. In the enormous living space provided by the sea, there are *plankton* (drifters), *nekton* (swimmers), and *benthos* (bottom-living organisms). Much of the benthos releases eggs and larvae into the plankton — this is the *meroplankton,* abundant in coastal waters. The larvae are dispersed by currents; they settle when their time has come, and start growing on the appropriate solid substrate. The meroplankton feeds planktonic and nektonic predators. Conversely, the plankton feeds the benthos. Thus there is an intimate ecologic relationship between free-swimming and bottom-living organisms. Ultimately, of course, benthic organisms rely on food produced in shallow waters, in the sunlit zone (Fig. 6.1). The notable exception is the deep-sea benthic community at the hot vents of the mid-ocean ridge (see Epilog).

Fig. 6.1. Spiny spider crab *(Paralithodes californiensis)* in La Jolla Canyon, at about 300 m depth. Note the sea grass which drifts in from the surf zone. Organic matter produced in the sunlit zone and transferred to depth provides food for benthic organisms. (Photo S. I. O.)

136

On the whole, fewer species developed in the ocean than on land. The living space of the ocean has fewer nooks and crannies than that on land, and the various parts of the world ocean are more interconnected than the land. Thus there is less opportunity for the isolation of populations in the ocean than on land, and hence less opportunity for separate evolution from populations to subspecies to species. The greater diversity of animals on land is entirely due to the proliferation of insects (>75% of one million animal species). Of the marine animal species (180,000) 98% are benthic; only 2% are planktonic or nektonic.

These numbers, of course, are subject to revision, by continuing investigations. The general trends seem secure, although we do not share the opinion of Pliny the Elder (23–79 A.D.) who stated, ". . . in the whole ocean, as large as it is, there lives nothing that we do not know."

6.1.2 Productivity and Environmental Factors. Sunlight and nutrients determine the growth and distribution of organisms, because the food chain has its base in the marine algae. Matter is transferred along this chain from primary production, to herbivores, to carnivores.

The algae need sunlight to grow (Fig. 6.2). The sunlit, or photic zone is only about 100 m thick. Below that depth little sunlight remains, perhaps 1% in very clear water. The angle of incidence of the sunlight, the cloud cover, and the amount of suspended matter in the water all influence the depth of penetration of the light. Benthic algae, then, can only occur on the upper half of the shelf area, say, over no more than 2% or 3% of the sea floor. Planktonic algae *(phytoplankton)*, of course, occur practically over the entire area. However, this does not mean that the production by benthic organisms can be neglected. Typically, phytoplankton production is of the order of 100 g carbon per square meter per year (100 g C/m^{2}yr). For benthic algae the production values can be up to one hundred times greater! Much benthic production takes place in salt marshes and in kelp forests, but also in the symbiotic algae of coral reefs.

While the ocean offers plenty of some of the materials necessary for growth, potassium and sulfate, for example, it is decidedly short of others such as phosphorus, fixed nitrogen, silica, and trace elements (iron, molybdenum etc.). These nutrients have especially low concentrations in surface waters, because they are constantly being removed by algae. In fact, their availability controls the growth of the algae; thus, they are *limiting nutrients*. Algal material, and also animal matter, sink below the photic zone and undergo bacterial decay. Nutrients are released in the process, that is, *remineralization* occurs. The nutrients are not being used at depth, in the absence of light. Hence, they become concentrated.

Thus, the ocean has a deep reservoir of nutrients below the surface waters. The boundary between this reservoir and the nutrient-poor sunlit waters is the top of the thermocline, usually between 50 m and 100 m deep.

When the thermocline is eroded by storms or when the deep water comes to the surface through *upwelling* (Sect. 4.3.3), the nutrients return to the surface water, into the sunlit zone. This fertilization by admixture of deep waters results in increased productivity, both in the algae and in the animals that feed on them. All the good fishing areas are in such regions of vertical mixing.

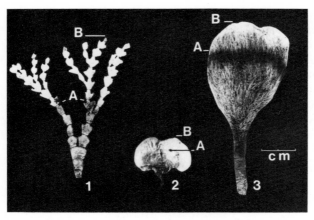

a

b

Fig. 6.2a, b. Sediment production and benthic photosynthesis.
a Production of algal carbonate, measured in the field by staining live specimens and observing growth till collection. *A* termination of staining; *B* collection.
1 Halimeda; 2 Padina; 3 Penicillus; all common benthic algae in Bermuda. (Photos and experiments by G. Wefer, 1980, Nature (London) 285: 323).
b *Acropora* meadow, Florida Keys. The staghorn corals produce abundant carbonate, with the aid of symbiotic algae (dinoflagellates) which live within their bodies. Growth rates are of the order of 1 cm/year (Photo W. H. B.)

6.1.3 Salinity. Life in the sea depends on other factors besides sunlight and nutrients. A large number of organisms can tolerate *salinity fluctuations* only if salinity stays between 30‰ and 40‰ (*stenohaline* forms). Examples are radiolarians, reef corals, cephalopods, brachiodops, and echinoderms. In general, their remains indicate marine conditions for the sediment that contains them. However, there are exceptions. A few representatives of the above groups may have relatively wide tolerances: for instance, there is a starfish that lives in the brackish waters of the Baltic sea. Also, fossils embedded in sedimentary layers do not necessarily show the environment of deposition: they may be redeposited.

With increasing salinity, for example in lagoons of arid regions, fewer and fewer organisms are able to survive until only a few representatives remain — an *impoverished* fauna and flora. Some ostracods (bi-valved crustaceans of millimeter size) can tolerate salinities of more than 100‰. As these hardy forms don't have many competitors (or predators), impoverished faunas can be very rich in individuals. Quite generally, of course, species-poor but individual-rich faunas indicate special, *restricted environments*. Restricted does not apply to space here, but to unusual temperature ranges, oxygen deficiencies, and other stress-producing factors.

6.1.4 Temperature. In the open ocean, where salinities are well within the tolerance of all marine organisms, temperature plays a decisive role in controlling distributions (Fig. 6.3).

In polar areas the temperature can fall to $-1.5°$ C; in marginal seas it can rise to well over $+30°$ C, as in the Red Sea and the Persian Gulf. Outside the tropics, temperature can vary strongly with the seasons, in the upper 100 to 200 m of the water column. Below this depth it varies little, however, being rather low throughout the year.

Most of the ocean is extremely cold: in the deep sea the temperature stays below $4°$ C. The cold temperature in itself does not reduce diversity: it has recently been shown that an astounding variety of small benthic organisms occur even at abyssal depths (Fig. 6.4). Large, unpredictable fluctuations of environmental factors appears to be much more restrictive than just extreme cold. Such fluctuations work damage especially on eggs, larvae, and the young recruits of a population.

Can we reconstruct the temperature distributions from a study of the remains of organisms? For open ocean conditions, such reconstruction has been honed to a fine art by the CLIMAP group (Climate Long-Range Investigation Mapping and Prediction) (Chap. 9). For shelf seas and enclosed basins the reconstructions are more difficult. Here, environmental factors besides temperature (salinity, muddiness of the water, seasonal bad weather) are much more important and unpredictable than in the open ocean. A change in a faunal assemblage, therefore, may be due to stress (or release of stress) in any one of these factors. One difficulty which exists for both deep sea and shallow sea, as far as paleotemperature estimates, is the *selective preservation* of faunal (and floral) assemblages (Sect. 8.3.2). Thus, changes in the assemblages may be due to changing conditions of preservation, rather than to changes in the conditions of survival and growth.

6.1.5 Oxygen. The content of *dissolved oxygen* in the water is another environmental factor of great importance, especially when concentrations fall to critically low levels of a fraction of one milliliter of gas per liter of water (normal: 4 to 7 ml/l). In cases where the oxygen concentration becomes low enough, all higher organisms, and even shell-bearing protists, will succumb, and only anaerobic bacteria remain. At the present time we have to go to certain special

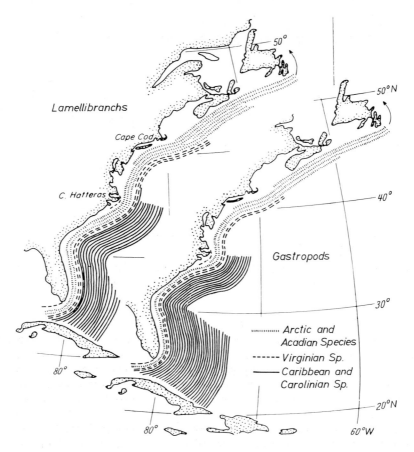

Fig. 6.3. Diversity of benthic shelf organisms as a function of latitude, shelf off the East Coast. An Arctic fauna with few species in the north is being replaced by species-rich temperate and tropical faunas to the south. (Each *line* equals 10 species). Note the sharp break at Cape Hatteras, where warm Southern waters (Gulf Stream) leave the shelf. Arctic waters penetrate down to Cape Cod. Gastropods show a greater change than bivalves (=lamellibranchs or pelecypods). The clams live mostly *within* the sediment, in contrast to the snails. Hence they are more protected from inclement conditions. (After A. G. Fischer, 1960, Evolution 14: 64; modified)

areas to study this phenomenon — fjords in Norway and Alaska, the Black Sea, the Santa Barbara Basin off California. In the geologic past, however, when the poles were not icy cold and did not therefore deliver oxygen-rich water to the deep ocean, conditions of oxygen deficiency were widespread. Much of the petroleum we burn today was formed in oceans with a low oxygen content, in regions where oxygen dropped below critical values and where organic matter did not readily decay, therefore.

Basically, oxygen deficiency arises when oxygen demand is strong, supply is weak. For example, in the Black Sea the salt water filling the basin (through the Bosporus, Sect. 4.3.5), comes from the Mediterranean and is covered with a layer

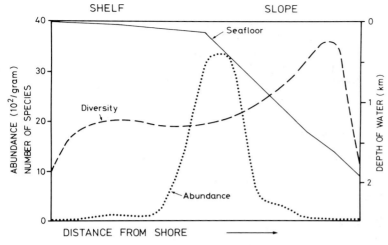

Fig. 6.4. Trends in abundance and diversity of benthic foraminifera across the shelf and slope off Central America. (After O. L. Bandy and R. Arnal 1957, Am Assoc Pet Geol Bull 41: 2037; simplified)

of fresh water brought in by Danube, Dnjepr, and other rivers. The light fresh water forms a lid on the heavy deep water, cutting off exchange with the atmosphere. Growth continues in the upper waters, delivering organic matter to the water below. Here it is eaten by animals, which use up oxygen by respiration, and is decayed by bacteria, also using oxygen. Thus, the deepest water becomes entirely anaerobic.

A somewhat analogous process can be observed in organic-rich muds. Free oxygen is only present in the uppermost layer of sediment, millimeter-thick (or centimeter-thick when sandy). Burrowing animals must set up air conditioning by pumping oxygen-rich water through their burrows — otherwise they must suffocate.

The geohistorian attempts to reconstruct the degree of oxygenation at the time and place of deposition of a given sediment layer, from clues such as lamination or nature of burrowing, and from chemical indicators such as sulfides and types of organic matter present. The remains of certain benthic organisms also provide information regarding the level of oxygenation (Fig. 6.5).

6.2 Benthic Life

6.2.1 Types. By far the greatest part of the sea floor is teeming with benthic organisms. Benthos which stays put is called *sessile*. All sponges, corals, brachiopods, and bryozoans are sessile. Benthos which moves about is termed *vagile*. It can move rapidly, like a startled crab, or slowly like the sluggish sea urchins, starfish, most bivalves, snails, and worms. Both groups have members

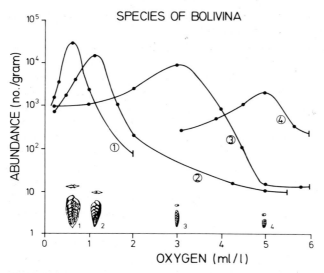

Fig. 6.5. Abundance of *Bolivina* species as a function of dissolved oxygen, California Borderland. *1 B. argentea; 2 B. spissa; 3 B. pacifica; 4 B. vaughani.* (From R. G. Douglas, 1979, SEPM Short Course 6: 21)

living *on* the floor or on top of other organisms (e.g., shells or kelp): *the epifauna.* Or they live hidden within rocks and sediment: *the infauna.*

Compared with the 125,000 marine species of epifauna there are only a paltry 30,000 species of infauna. Why? Are there more *niches* in the epifaunal way of life, that is, more opportunities for different approaches to making a living? We do not know. So far, biologists have not succeeded in determining the number of niches independently from the number of species occupying them.

6.2.2 Food and Substrate. Benthic animals, ultimately, live on food stuff falling down through the water or coming in from upslope along the sea floor. Both living plankton and dead *detritus* drift in the water: the *seston.* The detritus consists of organic and inorganic particles. One consequence of the dependence of the benthos on this "rain" of seston is a pronounced decrease in benthic biomass with depth: the deeper the sea floor and the farther away from the fertile coastal zone, the smaller the amount of nutritious rain reaching it. Although so very little reaches the abyss (order of 1% of the production) hundreds of species — tiny crustaceans and worms — make their living from the scraps coming down. How do these organisms manage to spend less energy in the search of food than the food is worth? They mus be extremely energy-efficient.

The *sessile benthos,* of course, does not seek out its food. It waits for the water to bring suspended material from which to take nourishment. The *suspension feeders* filter the water, either passively, using the natural water flow, or actively, moving water past their straining apparatus. *Commensals* may seek a free meal in addition to shelter: for foraminifera living in sponges it is convenient that the host

142

provides both protection from enemies and brings in a steady stream of food particles. Sessile benthos is especially abundant on rocky bottom.

The *vagile benthos* on the rocky substrate — starfish, sea urchins, gastropods, ostracods — feeds on epibenthic organisms, for example, by scraping off algae, or preying on sessile animals. It protects itself from storms and predators by hiding in nooks and crannies, growing thick shells, or by clinging to the rock as do the chitons and patellas using their strong sucker foot. On soft bottom, most vagile benthic animals ingest sediment *(deposit feeders)*. Others hunt for prey.

We see that life on the bottom is greatly influenced by the type of the substrate, which like the organisms themselves reflects environmental factors: temperature, salinity, oxygen content, currents, microtopography. Thus, sedimentary processes and life on the sea floor are intimately associated. The remains of organisms within their habitat yield valuable clues to conditions, of both growth and sedimentation therefore. In fact, benthic organisms may *produce* much or all of the substrate.

One aspect of ecology which is of great interest to geology is the *production rate* of hard parts, that is, carbonate (Fig. 6.2). Off Miami the macrobenthos produces annually about 1000 g carbonate per m^2 in the tidal zone, between one and 400 g per m^2 in the deeper water offshore. On shallow areas of the Persian Gulf a single species of foraminifer, *Heterostegina depressa* (Fig. 6.6), delivers annually 150 g carbonate per m^2. This protist lives in symbiosis with photosynthesizing algae, as do corals. Questions regarding type and rate of mineralization are obviously of importance to paleoecology; here biological and geological research are closely intertwined. Benthic foraminifera whose shells provide a lasting record are ideally suited to define the limits of many of these conditions (Fig. 6.6, also see Figs. 6.4 and 6.5).

For an appreciation of the relationships between organisms and environment of deposition, let us have a closer look at bethic organisms and their substrates, hard and soft.

6.3 Organisms and Rocky Substrate

Rocks on the ocean floor crop out where currents or waves keep it free from sediments, or in areas of active erosion. Steep-sided walls in fracture zones and other rugged topography also are likely places for rock outcrops.

A great number of different materials can be called rock: ancient sandstones and limestones of wave-cut platforms or along submarine canyons, basalt scarps or manganese pavements, submerged dead coral and cemented beach sand (beach rock), boulders dropped from icebergs or left on the shelf in moraines, sunken ships, and other man-made objects.

The most abundant type of rocky substrate undoubtedly is the pillow basalts of the Mid Ocean Ridge. These basalt outcrops are commonly rather barren, except in local spots of hydrothermal exhalation, in the Pacific. Here, a rich fauna of bivalves and other sessile and vagile benthos has been discovered recently (see Epilog).

143

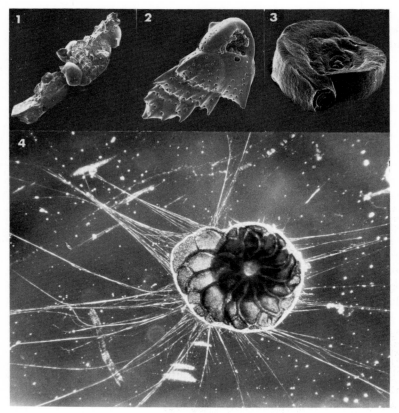

Fig. 6.6. Examples for different types of foraminifera.
1 Arenaceous form which agglutinates available particles (*Reophax,* x 50); *2, 3* calcareous forms (*Trimosina,* x 250; *Spiroloculina,* x 90). The species of *Trimosina* shown lives at a depth of greater than 20 to 30 m, in the Indian Ocean; *Spiroloculina* prefers shallow water and coarse substrate. *4 Heterostegina depressa* (diameter of test: 0.84 mm) a large tropical form living in shallow water in symbiosis with photosynthesizing algae. The "pseudopods" radiating from the test are used for anchor and also for locomotion. (SEM photos C. Samtleben and I. Seibold. Microphoto R. Röttger)

In shallow areas, rocky bottoms are commonly covered by algae. The algae can be microscopic diatoms (many of these move about on the rock), or up to several-meters-long soft ribbons waving in the currents, offering hideouts and hunting grounds for fishes. Hence rocky areas are good fishing grounds (although they can be hard on the fisherman's nets).

Encrusting algae tend to cover rock outcrops permanently. Depending on water temperature and depth, *epifaunal* associations develop: sponges, corals, tubeworms, oysters, barnacles, bryozoans, encrusting foraminifera (Fig. 6.7). The *infauna* bores actively into the substrate. This remarkable ability has been acquired by many types of organisms, including sponges, worms, and molluscs. Some sea urchins with unpleasantly (for swimmers) long and brittle spines live in custom-made cavities. At first glance such burrowed rock gives the appearance of being quite solid, since the entrances to the burrows are usually quite small. However, in reality the outer rind of the rock may resemble Swiss cheese in its structure. Storm surf can then destroy this outer layer. Thus the rock wears away largely through bio-erosion. Limestone is especially susceptible to this type of attack. Debris from boring sponges grinding holes into calcareous rock can make up some 30% of the sediment in certain lagoons of the Pacific coast.

Fig. 6.7. Dense epifaunal growth. **1** Edge of coral patch reef, Lizard Island, Great Barrier Reef (14°45′N, 2 m). (Photo E. S.). **2** Epifauna on a mound on the Ross Sea shelf (Antarctic, 76°59′S, 167°36′E). No algae are present at this depth (110 m) so that the dense growth here is all from animals: large fingery sponges, small bushy bryozoans, finely branched horny corals. A sea lily (echinoderm) is seen in the *upper left*. (Photo J. S. Bullivant, 1967, N Z Dep Sci Ind Res geol Surv Paleontol Bull 176)

145

In shallow rocky areas, organisms commonly are removed periodically by wave action, and their remains collect in depressions or at the foot of the rocks. The *reef-talus* deposits are of this origin. They are coarsely layered thick sequences of sediments surrounding the reef structure, and are familiar from ancient exposed reefs on land (Permian El Capitan Reef, west Texas; Triassic Dolomites in the Alps; Jurassic Malm reefs in middle Europe; Cretaceous Albian reefs in southern Arizona). The coarse-grained material of the reef talus is highly porous and permeable, and can serve as reservoir rock for petroleum — as in the oil fields of the Middle East.

We shall return to reef formations in Chapter 7 when discussing the significance of marine deposits as climate indicators.

6.4 Sandy Substrate

Rocky ground quite commonly has a rich, "flashy" assemblage of diverse organisms, from tropical reefs to Antarctic sponge forests (Fig. 6.7). Substrate of pebbles and boulders is much like rocky bottom provided there is no movement. If waves move the rocks, most sessile benthos cannot survive.

The benthic communities on sandy sea floor are much less flashy. Commonly one sees rather little — the organisms are mostly hidden within the sand. The reason why sessile epifauna is largely absent is the instability of the substrate. Where it is stabile, sea grass can take hold and further stabilize the ground. Epifaunal diatoms, foraminifera, and bryozoans grow on such grass.

Vagile benthos, with or without burrows to return to, can be quite abundant on the sandy bottom. Crabs and snails are common sights. In the intertidal range there is a large variety of nonmarine invaders during low tide, mainly birds. These hunt for the hidden infauna. Conversely, during high tide the invasion is from the sea: sting-rays and other fishes digging up worms and molluscs.

The presence of so many predators, and the shifting of the sand which can suddenly expose the infauna, require the ability to burrow very rapidly. Crabs demonstrate this adaptation very obviously, but also many clams can burrow quickly — as clam diggers well know. Burrowing clams have a strong long foot which they extend into the sand below, then inflate by water pressure to anchor it, and pull the rest of the body down through muscle contraction. A smooth outer shell, usually quite sturdy, characterizes such clams. The study of burrowing mechanisms and the resulting tracks in the sediment is a research field by itself (Sect. 6.6).

Burrowing clams are *suspension feeders,* with an inhaling and an exhaling siphon (Fig. 6.8). After death, the shells of such bivalves — hydraulically quite different from the surrounding sand — are sorted out and concentrated into layers of coquina (Fig. 6.9). Such coquina deposits are widespread in the geologic record and can be used as marker beds locally.

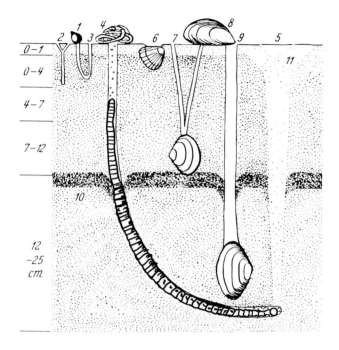

Fig. 6.8. Sediment reworking by benthic organisms in tidal flats of temperate regions. *1* Snail *Littorina; 2* worm-burrow *(Pygospio), 3* crustacean burrow *(Corophium), 4* and *5* worm *Arenicola* with fecal mound and funnel (see photo below), *6* to *9,* bivalves *Cardium, Scrobicularia, Mytilus,* and *Mya; 10* originally horizontal layer, disturbed by burrowing, *11* light brown surface sediment. (After H. M. Thamdrup, in R. Brinkmann (ed), Lehrbuch der allgemeinen Geologie. F. Enke, Stuttgart 1964; modified. Photo E. S.)

6.5 Muddy Bottom Substrate

Muddy and clayey substrates present different kinds of challenges for their benthic denizens. Thus the faunal assemblage of this habitat differs greatly from those of the sandy and rocky environments. Muddy substrates are rich in inorganic and organic particles of extremely small size — less than one to a few microns

147

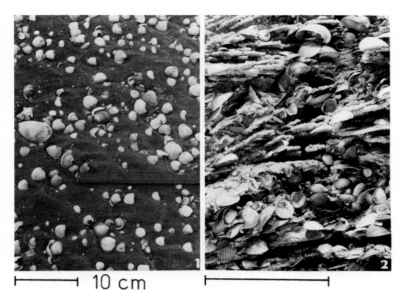

├───────┤ 10 cm

Fig. 6.9. Shell pavement in the upper intertidal, German North Sea (N of Weser estuary). The shells form low mounds, built by waves and tidal currents. The orientation of the shells on the sea floor indicates current directions [note valves pointing to right in **1** and shingling in box corer profiles in **2**. (Photos E. Seibold **1** and H. E. Reineck **2**)]

(1 micron = 0.001 mm). This has two consequences. First, the substrate is rather less shifting than the sand, due to the high cohesion of the clay particles. It is more difficult to burrow through than sand, but burrows have a better chance to persist. Second, the high content of organic matter makes it worthwhile for many organisms to pass it through their guts and extract the digestible fraction. Hence, *deposit feeders* are commonly found on and in this type of substrate. Of course, *suspension feeders* would thrive on the organic matter also. However, their filtering apparatuses tend to get clogged with the abundant (and to them useless) clay. Furthermore, by constantly reworking the mud, the deposit feeders make it difficult for sessile benthos to find a stable foothold. Thus, deposit-feeding benthos dominates: about three fourths of the benthos belongs to this group.

Deposit feeding can take place on top of the sediment, which may also yield some diatoms and other algae growing there. Snails typically represent this type of feeding behavior, for example *Littorina* in tidal flats (Fig. 6.8). Also, burrowing clams may extend their feeding siphons out to the top, to pipette off the food on the surface. Examples are *Scrobicularia* and *Macoma*. On the tidal flats of northwestern America, the small species *Macoma secta,* with a length of only 6 to 7 cm, can extend its siphon up to 1 m — covering a good-sized territory. Various types of worms, and sea cucumbers (holothurians) are deposit feeders sometimes seen in deep sea photographs. They leave tracks and fecal strings on the surface, burrows within the sediment (Fig. 6.10). In shallow water certain deposit feeders can become extremely abundant. The lugworm *(Arenicola),* typical for muddy

148

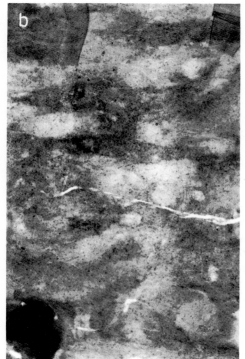

Fig. 6.10. Tracks and burrows in deep sea clay.

Upper: furrow-like feeding trails of holothurians (deposit feeders). Note range of states of preservation, fresh trail at *lower left.* Also meander-like fecal strings extruded onto sediment surface. Small compact structures probably are *agglutinating foraminifera.* Tracks and trails are not preserved in the record, unless burial is by sudden influx of sediment (such as turbidites). Northern Solomon Trench, water depth 8500 m, field of view in upper photo 1.5 m × 3 m. (Photo courtesy R. L. Fisher.)

Lower: X-ray radiograph, positive, of deep sea Red Clay. Dense areas are dark. Network of burrows (mainly deep-reaching burrows are preserved) illustrates intensity of bioturbation. Typical burrow diameter, 0.5 to 1 cm. Lower *left:* small manganese nodule. Core taken about 1000 km SE of Hawaii, water depth 5000 m, 49–60 cm below sediment surface. (Photo F. C. Kögler, Kiel)

intertidals in many areas can attain densities of 200 individuals per square meter (Fig. 6.8). Each worm can ingest several hundred grams of sediment per day. At high densities, the entire sediment, down to about 20 cm depth, may be worked over in a matter of weeks, by deposit-feeding organisms.

149

Muddy bottoms, then, are largely fecal material, many times recycled. This is also true for the deep sea floor, although here the cycling takes thousands of years rather than months as in the mudflats — a difference of a factor of 10,000. Of course sedimentation rates also differ considerably, by a factor of up to 1,000. The greater number of recycles in the shallow water, before burial, reflects the much higher supply of organic carbon — and hence energy supply — to the coastal sea floor.

If fine sediments are fecal matter, should one not see more evidence of fecal pellets and fecal strings? Indeed, fecal pellets are extremely abundant, especially toward the surface of muddy sea floor. Here they alter entirely the hydraulic properties of the sediment, give rise to bacterial growth, and to development of fungal mats in places. In the case of calcareous mud, the fecal pellets can harden by cementation and can thus fossilize more readily. They are very abundant in many limestones, as near-spherical and oval grains of diameters between 0.03 and 0.1 mm, a fact that is only realized after careful microscopic study.

The biological reworking of soft sediment on the sea floor is a process of global geochemical significance. Without such *bioturbation,* the sediment would quickly disappear from the marine chemical system. Only a thin upper layer would readily react with the sea water. Through bioturbation, however, a several centimeter thick layer of sediment keeps exchanging matter with sea water, the falling organic material remains available for some time before burial, the nutrients in it are remobilized and are given back to the seawater. Thus, the overall *fertility of the ocean* is closely linked to bioturbation.

We can take another step — a bold one — and speculate that the *climate of Earth* is moderated by bioturbation processes. How might this work? Consider the carbon dioxide in the atmosphere (0.03%). It is generally believed to be closely linked to the global temperature, through the "greenhouse effect". It is transparent to the incoming radiation of the sun but tends to absorb infra-red radiation attempting to escape into space. Clearly, CO_2 is involved in biological cycling, hence in the fertility of the ocean. That is one link to bioturbation. Another is more subtle: if CO_2 input by volcanoes were to increase suddenly, the ocean would presumably react by taking up CO_2 and by subsequently dissolving carbonate from the sea floor, especially the deep sea floor. Bioturbation would greatly facilitate such a reaction. Thus, bioturbation should help to normalize CO_2 concentrations in the atmosphere — a feedback that is of considerable interest in view of the present high CO_2 production by burning of fuel (see Sect. 8.3.6).

6.6 Trails and Burrows

6.6.1 Trace Fossils. In the foregoing, we have looked at sea floor and organisms with an appreciation for the biological viewpoint. However, the geologist's ultimate concern is the *record:* how is it made, what can it tell us about conditions of the past. An entire branch of geology — *ichnology* — has grown from the study of the tracks, trails, burrows, and other sedimentary disturbances made by

150

organisms. We next turn to some of the problems arising in the study of such trace fossils. We also have to ask just how bioturbation disturbs the orderly recording of events in deep sea sediments — the sediments from which we hope to extract the most complete and accurate information about ancient climates (as discussed in Chap. 9).

There is a great variety of traces, which eventually become trace fossils: trails, feeding tracks, and fecal strings on the surface, burrows stuffed with fecal mud, burrows used for housing and later filled by washed-in sediment, and others (Fig. 6.11). Various kinds of worms, snails, bivalves, crabs, holothurians, sea urchins, and star fish make such tracks and burrows. Crabs and shrimp make the longest and deepest burrows. In certain cores from off Northwest Africa we have observed vertical burrows more than 3 m long! Such "lebens-spuren" have been, and are being, intensively studied by marine geologists in Wilhemshaven on the North Sea, in Kiel, and in Tübingen. Thus, the German word *Lebensspuren* which translates into *life-traces* has been adopted into English usage.

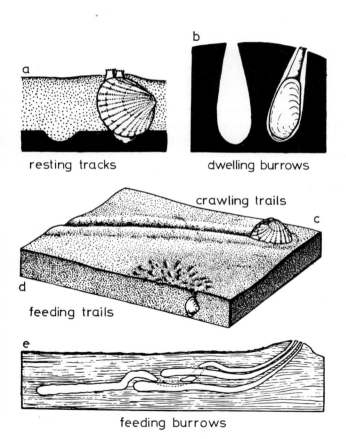

Fig. 6.11a–e. Types of traces by function, and organisms producing them. **a–d** bivalves, **e** worms. (From A. Seilacher, 1953, Neues Jahrb Geol Palaeontol Abh 96: 421 and 98: 87)

151

6.6.2 Lebensspuren are, in a sense, fossilized behavior of benthic organisms. Much like shells, behavior represents adaptation to the environment, hence we can read environmental information from tracks and burrows. Unlike shells, lebensspuren cannot be transported by currents, but stay in the sediment where they were made. Thus they indicate in situ conditions, whether it be food supply, stability of the sediment, wave action, even depth of water.

The tracks and burrows are not only useful for paleoecologic reconstruction, but also for physical geology, and even for tectonics. In the deep sea, for example, vertical burrows do not last if the sediment has a tendency to shear horizontally. Through systematic box coring of deep sea carbonates in recent years it was found that vertical burrows are extremely abundant in the west equatorial Pacific, in shallow areas with sandy, stiff sediment. In deep areas, where dissolution of carbonate breaks down the sand, the sediment loses strength. It can flow when shaken by earthquakes. There are no vertical burrows here, only horizontal ones.

On land, the field geologist sometimes deals with badly jumbled marine strata — he may not even be able to tell which side of a layer is up! Burrows, if present, can help decide the issue. Many burrows are U-shaped. If one finds them as an arch, that is, upside down, it proves that the layer has been inverted by tectonic forces. Sometimes there is a question of *lacustrine* (lake-) versus lagoonal or marine sediments. The marine environment has large long burrows in contrast to the less diverse and generally smaller ones of lakes.

6.6.3 Preservation. Not all tracks and burrows are equally well-preserved, of course. A delicate surface trail made by a star fish or a mussel has little chance of entering the record. A 30-cm-deep *Arenicola* or *Mya* burrow or especially a 2-m-deep crab burrow has an excellent chance: we should have to remove a thick layer of sediment in order to obliterate it. In general, surface tracks can only be preserved by a catastrophic deposition event as by a flood or a turbidity flow. In the case of continuous accumulation, when sediment builds up gradually and there is a mixed layer from bioturbation, surface tracks are destroyed while only burrows penetrating the mixed layer are preserved in the record (Fig. 6.12).

6.7 Bioturbation

6.7.1 Effects of Mixing. Bioturbation prevents the preservation of thin layers, that is, annual layers, thin turbidite layers or layers produced by contour currents alike. Thus, today's well-turbated slope sediments are poorly layered. However, this need not always have been the case. During times when the ocean was not well oxygenated, burrowing organisms could not as readily do their work of homogenizing the incoming sporadic supply of sediment. Thus, quite commonly, we see finely layered slope sediments in the geologic record, deposited in warm oceans with low oxygen content. (Warm water dissolves much less oxygen than cold water: only since the Earth has grown ice caps is the deep ocean highly aerated.)

152

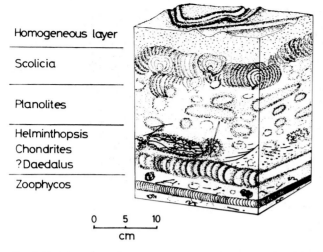

Homogeneous layer

Scolicia

Planolites

Helminthopsis
Chondrites
?Daedalus

Zoophycos

0 5 10
cm

Fig. 6.12. Biogenic sedimentary structures in deep water off NW Africa (about 2,000 m). Tiered arrangement of active burrowers reaches 30 cm into the sediment. Bioturbation homogenizes the uppermost 3 cm of sediment completely, destroying surface tracks. Sea urchins (*Scolicia* burrows) and other burrowers produce typical traces (*Planolites* etc.) in different depths within the sediment. (After A. Wetzel, 1979, Ph D Thesis, Geol Inst Kiel)

Bioturbation alters and smoothes the record. In a finely layered sediment, each layer has a message about its origin and its environment of deposition. But when the layers are mixed the messages are mixed also, and we obtain a signal which is some sort of an average of the conditions which existed at the time. How large a time interval is being averaged? How does this averaging affect the *dating* of sediments? These questions are being investigated on deep sea box cores, because of the great importance of the deep sea record for detailed paleoclimatic reconstruction. (For dating by radioisotopes see Appendix A 8.)

One way to proceed is to determine the exact concentration of radioactive carbon, as a function of depth in the sediment. The ^{14}C-stratigraphy should give us a clue to both the depth of mixing and the effect on age determination (Fig. 6.13).

The ^{14}C enters the sedimentary record within the $CaCO_3$ of the carbonate shells. A certain proportion of the CO_2 in the air (and hence of the HCO_3^- in the surface water) contains ^{14}C, which is produced in the atmosphere through the activity of cosmic rays, from the normal nitrogen-14 atom. The radiocarbon is incorporated into living matter and shells. It decays back to nitrogen. Thus, young shells have more radiocarbon than old ones. The decay rate is such that one-half of the C-14 is gone after 5,700 years. This is the *half-life*. Thus, we can predict that a shell will retain one half its radiocarbon after 5,700 years, one fourth in 11,400 years, one eighth in 17,100 years, and so on. The limit of measurement is near 35,000 years — one sixty-fourth of the original concentration.

How do we know the initial concentration of radiocarbon? We must assume that is was the same then as it is today in freshly forming shells. Using this

153

assumption, we can calculate the *apparent age distribution* down the core. This is the distribution produced both by the continuous sedimentation of radioactive sediment (i.e., calcareous shells) *and* the mixing on the sea floor. There is an interesting trend in the ages: they change very little in the uppermost part of the sediment, the *mixed layer,* and then show a regular progression to higher values downcore.

6.7.2 Mixing Model. Sediment mixing is a complicated process involving the burrowing activities of various types of organisms, disturbing the sediment to various depths (Fig. 6.12). It is not yet possible to describe the physics of mixing in a way which is both correct and useful. However, the pattern shown by the ^{14}C-distribution suggests a very simple (somewhat too simple, for sure) model of how mixing operates on sediment, as follows.

We assume that the sediment consists of two layers only: a mixed layer at the very top and a historical layer below. At the top of the mixed layer the newly arriving material builds up the sea floor at a rate

$$s = dl/dt$$

where s is the *sedimentation rate,* and dl is the difference in height produced within the time interval dt. At the bottom of the mixed layer an equal amount of sediment leaves, per unit time, and enters the historical layer. The mathematical formulation of this model is straightforward. For example, a tracer such as microtectites (from a meteor impact) or volcanic ash, which is added to the sea surface only once, will show a distribution following the decay equation:

$$C_L = C_o \cdot e^{-L/M} \tag{6.1}$$

where C_L is the concentration of the tracer at a distance L above the first occurrence, C_o is the original concentration in the mixed layer, right after introduction of the tracer, and M is the thickness of the mixed layer. Application of this formula to the distribution of volcanic ash, observed in cores from the North Atlantic, suggests that M is in the neighborhood of 6 cm. This estimate agrees well with the mixed layer thickness suggested by the ^{14}C-stratigraphy shown in Fig. 6.13, for a core from the west-equatorial Pacific.

6.8 Limits of Paleoecologic Reconstruction

We have discussed environmental factors, especially physical ones, and we have looked at the production of lebensspuren and what they may tell us about the environment. Also, we have introduced the problem of bioturbation, the process which alters and smoothes the record. What about the organisms themselves, their interactions, their reproduction rates, their survival strategies, their larval dispersion? How do larvae know where to settle? How many survive? Who

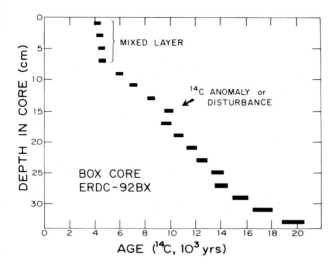

Fig. 6.13. Carbon-14 stratigraphy of a box core from the western equatorial Pacific. *Top layer* shows similar ages down to 7 or 8 cm, due to mixing. Overall sedimentation rate is near 1.7 cm/1000 yrs. An anomalous ¹⁴C sequence appears at 15–17 cm depth, presumably caused by a redeposition event. (Data in Peng T-H et al., 1979, Quat Res 11: 141)

competes with whom? Who eats whom? Which are the symbiotic relationships, which parasitic? How do they control distributions? And especially: how much of all this information enters the geologic record?

These problems are constantly encountered when studying communities, here, *benthic communities*. Already the definition of communities is difficult — the biologists use "organisms that are often found together". Presumably, their web of interactions provides the checks and balances which keeps a community *stable,* that is, the same kinds of organisms in roughly the same proportions live together over extended periods of time.

The *record* also has "communities", that is, co-occurring fossils. Not surprisingly, the difficulties in guessing the *interactions* of organisms long dead, are virtually insurmountable.

Even seemingly simple questions are hard to answer. For example, how large a territory does a benthic foraminifera feed on? On the shelf of the Arctic there are up to 50 individuals per square meter of a foraminifer of the genus *Astrorhiza*. Their shells are about 5 mm in diameter. But when alive, the net of pseudopodia takes up an area 6 cm in diamater, an area more than 100 times greater than that of the shell. A large part of the sea floor may be "occupied" by these tiny organisms. How would we know this if we only had their shells?

It does not help, of course, that only a selected portion of the community is eventually fossilized. Let us inspect some shallow water communities recognized especially through studies of Danish biologists (Fig. 6.14). In many Arctic and cool-temperate regions the *Venus* community lies offshore in 10 to 20 m water depth. *Venus,* a bivalve, is a conspicuous member. Others are the star fish

155

Fig. 6.14a, b. Typical infauna associations in shallow water in cold to temperate regions. Examples are from the entrance to the Baltic Sea (Kattegat). **a** *Venus* Association in sandy substrate at 10–20 m water depth, with the sea star *Astropecten 1*, the sea urchin *Echinocardium cordatum 2*, the bivalves *Venus gallina 3*, *Spisula elliptica 4*, *Tellina fabula 5*, the polychaete tube-building worm *Pectinaria koreni 6*, and the gastropod *Natica 7*. *b Amphiura* Association in muddy substrate at about 20 m water depth with the brittle stars *A. chiajei 1* and *A.filiformis 2*, the snails *Turitella communis 3* and *Aporrhais pespelicani 4*, the sea urchin *Brissopsis lyrifera 5* and the polychaete *Nephthys 6*. Coastal regions in other regions of the world have other species, but the basic make-up of the associations persists ("parallel communities"). (After G. Thorson, 1972, Erforschung des Meeres, Kindler, München)

Astropecten, the sea urchins *Echinocardium*, and the polychaete tube-building worm *Pectinaria koreni*. Species may change, but the genera and the structure of the community are pretty much the same everywhere *(parallel communities)*. As geologists we can expect the preservation of *some* of the shells, *some* of the burrows. Nothing recognizable will be left of the worms or of the shrimps. Nothing also of the predators which invade the area from the sea and from the air.

Muddy sand and deeper waters are settled by the *Syndosmya* community. The name-giving clam is a favored food item of flat-fish — something that certainly would be hard to tell from the fossil record.

Muddy bottoms in depths of about 20 m are conspicuous by their dense cover of brittle stars (*Amphiura* community, Fig. 6.15). Up to 500 individuals lie on a square meter. They dominate the ecology of the sea floor here, but it is doubtful that this information would be preserved in the record.

Fig. 6.15. Brittle star community, Kiel Bay, ca. 20 m depth. (Photo Diving Group, Geol Inst Kiel)

7 Imprint of Climatic Zonation on Marine Sediments

7.1 Overall Zonation and Main Factors

7.1.1 The Zones. The chief factor in producing climatic zonation is the amount of energy received from the sun — it is high in the tropics, low at the poles (Fig. 7.1). A coarsely latitudinal zonation of the oceans generally employs the categories *tropical, subtropical, temperate,* and *polar,* whereby the poleward part of temperate and the more temperate part of polar could be distinguished as *subpolar* (Fig. 7.2).

The *tropical zone* has an excess of heat, which it exports. Seasonal fluctuations are minimal; average temperatures are near 25° C, with sustained open ocean maxima close to 30° C. Near the Equator, daily rainfall, cloud cover, and weak winds, lead to excess precipitation over evaporation. At the Equator proper ($\pm\,2°$ latitude) fertility is high because of equatorial upwelling. Elsewhere it is low, except near continents.

The *subtropical zone* is the broad region between the tropics proper and the temperate areas. It is the desert belt, both on land and in the sea. Cloud cover is low, evaporation rates are high, and salinities, therefore, attain values well above average. This zone is invaded by the neighboring climate regimes depending on seasons. Annual temperature ranges can be very high. Coastal areas show seasonal upwelling, depending on wind strength.

The *temperate zone* is strongly influenced by seasonal change. Rainfall generally exceeds evaporation, and salinities are correspondingly reduced. Being the transition between the warm and the cold regions of the planet, the temperate zone has strong temperature gradients, hence strong winds. These force mixing of surface waters with (nutrient-rich) waters in the upper thermocline, along the west-wind drift. The temperate zones are fertile regions, therefore. In the poleward parts of the temperate zones mixing is further enhanced by seasonal break-down of the thermocline.

The *polar areas,* finally, take up the smallest portion of the globe, but they are of prime importance as makers of climate. Their ice rim fixes the endpoint of the overall temperature gradient, which ultimately controls winds, currents, and evaporation-precipitation patterns.

The climatic zones are not exactly parallel to latitudes: note how the boundaries are shifted by the currents of the subtropical gyres, especially the Gulf Stream and its west-wind extension.

7.1.2 Temperature and Fertility are the most important climatic factors in the ocean, as far as the production and distribution of biogenous sediments. In the

158

Fig. 7.1a, b. Climatic extremes in the present ocean. **a** Fringing reef and lagoons, Huahine, Society Islands, tropical South Pacific. (Courtesy D. L. Eicher). **b** Tabular iceberg, Antarctic Ocean. (Courtesy T. Foster)

tropics and subtropics nutrients are limiting to growth. Here the surface waters become depleted of nutrients because the density gradient (light warm water on top of cold heavy water) hinders upward diffusion of nutrient-rich deep waters. Generally, therefore, production is low. In high polar areas light is the limiting factor: productivity also is low. In the coastal zones of the tropics, and in temperate latitudes the supply of light and of nutrients both is adequate; here production is high (Fig. 7.3).

159

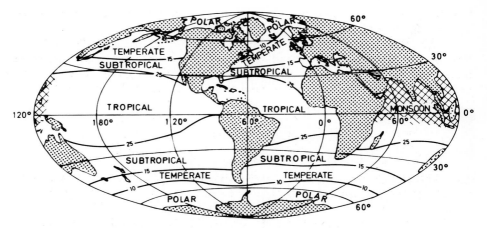

Fig. 7.2. Climatic zonation of the open oceans. Zone boundaries tend to follow latitudes and line up with climatic belts on land. Temperature, seasonality, and water budget (evaporation-precipitation balance) are the most important descriptors. Temperate and polar can be separated by another zone: subpolar. Approximate temperatures of surface waters in °C shown at the boundaries

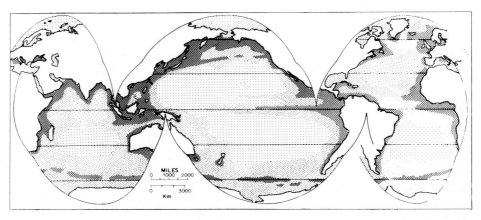

Fig. 7.3. Primary production in the world ocean (in mgC per m² per day). (From W. H. Berger, in J. P. Riley, R. Chester [eds], 1976, Chemical oceanography, vol V, 2nd edn, Academic Press, London, New York; based on a compilation of O. J. Koblentz-Mishke et al. in W. S. Wooster, 1970, Scientific exploration of the South Pacific, Natl Acad Sci USA)

Nutrient supply, as already mentioned, is increased through vertical mixing. Upwelling areas, where water from the thermocline reaches the surface, straddle the coastline in many regions — off California, off Peru, off western Africa (Sect. 4.3.3). Such upwelling is largely seasonal. On the continents, seasonal contrasts are exaggerated, compared with the oceans (continental climate). Around the large and high land masses of East Asia this seasonal contrast affects the entire northern Indian Ocean (monsoonal climate, see Fig. 7.2, hachured area). During winter, cold air masses flow off the continent (Northeast monsoon in India)

160

bringing drought and upwelling. During summer, the circulation is reversed: the Southeast monsoon brings rain from the ocean.

Biogenous sediments are excellent indicators of climate: many organisms have narrow ranges of temperature and fertility and the presence of their remains immediately yields excellent clues for climate reconstruction. Also, the chemistry of the shells can give indications of temperature, salinity, and growth rate.

7.2 Biogeographic Climate Indicators

7.2.1 Paleotemperature from Transfer Methods. The most widespread biogenous deposit on the present sea floor is calcareous ooze, which consists of plankton shells.

Of the shell-making plankton, the foraminifera were found to be the climate indicators par excellence. They are reasonably diverse, widespread, and easily identified under a low-power microscope (that is, after some practice). Planktonic foraminifera live mainly in the uppermost 200 m of the water column. There are about 20 abundant species, and as many rare ones. Each climatic zone has one or more characteristic species and a typical abundance pattern of species within it. Some temperature-sensitive species are shown in Fig. 7.4, with the areas they characterize. The general agreement of the faunal zonation with the climate zonation of Fig. 7.2 is obvious — the role of latitudinal bands and of the deflection by surface currents again is apparent.

7.2.2 Transfer Equations. The agreement between faunal and climatic zonation suggests that it should be possible to reconstruct climate by *counting the relative abundance* of the species in the sediment assemblages. This is indeed the method most commonly applied. The counts are used to calculate the original surface water temperatures by means of a *transfer equation,* a term coined by the paleoecologist John Imbrie. To illustrate how this works let us consider a very simple transfer equation, namely the weighted average of the optimum temperature:

$$T_{est} = \Sigma p_i \cdot t_i / \Sigma pi \qquad\qquad (7.1)$$

Here T_{est} is a temperature estimate based on an assemblage of foraminifera (or other shells), p_i is the proportion of the i-th species and t_i its temperature optimum. The optimum, in this context, is the temperature at which we find the highest proportion of the species in the calibration set.

The *calibration set* consists of sea floor surface samples and corresponding surface water temperatures. For example, we may find that *on the sea floor* species no. 3 has its highest proportion underneath surface waters with an annual average temperature of 20° C. We call this its optimum, and designate it as t_3. For the temperature estimate of a given sample, the product $p_3 \cdot t_3$ is one of the terms of the above summation. The larger the p_3, the closer the sum will be to 20° C. The

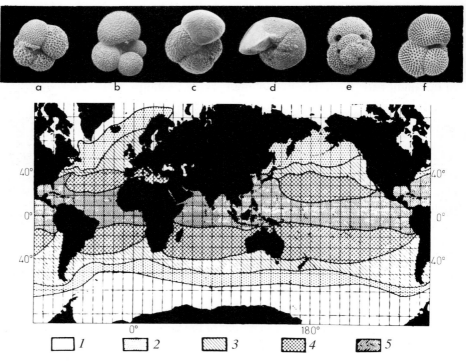

Fig. 7.4. Major distributional zonation of planktonic foraminifera. *1* Arctic and Antarctic; *2* subarctic and subantarctic; *3* transition zone; *4* subtropical; *5* tropical. Dominant foraminifera: *a Neoglobo-quadrina pachyderma. zone 1* left-coiling, *zone 2* right-coiling. *b Globigerina bulloides, Zone 2. c Globorotalia inflata, Zone 3, d Globorotalia truncatulinoides, Zone 4. e Globigerinoides ruber, Zones 4* and *5. f Globigerinoides sacculifer, Zone 5.* (Map from A. W. H. Bé, D. S. Tolderlund, 1971, in B. M. Funnell, W. R. Riedel [ed], 1971, The micropalaeontology of oceans. Univ Press, Cambridge. SEM photos courtesy U. Pflaumann, Kiel)

presence of species with a higher optimum temperature will pull the result up above 20° C, and vice versa. Under favorable conditions, this simple equation will estimate the correct temperature to within better than 3° C. The technique can be improved by considering the *temperature range* of each species, and by regression of the estimated temperature on the expected temperature, in the calibration set. The method then gives excellent results. More powerful transfer techniques also exist, based on the same principle of calibration. Factor-regression analysis is one of these techniques. It is used by the CLIMAP group to map conditions in the ice age ocean.

7.2.3 Application of Transfer Methods: Climatic Transgression. Great progress has been made in determining the temperatures of the sea surface for the latest Pleistocene, especially in the North Atlantic. The information is retrieved from numerous sediment cores distributed throughout the region. An example of such reconstruction, by the CLIMAP group, is given in Fig. 7.5.

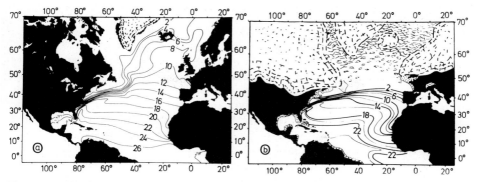

Fig. 7.5a, b. Expansion of polar climate in the North Atlantic, 17,000 years ago. **a** Today's sea surface temperature in winter. **b** Transfer map, winter surface temperature 17,000 years ago (in ocean), and distribution of ice sheets and pack ice. (Data from CLIMAP in E. Seibold, 1975, Naturwissenschaften 62: 321; see also Science 191: 1131, 1976; Geol Soc Am Mem 145: 464 pp, 1976)

The distribution of surface water temperatures — and hence of surface currents and water masses — differs greatly for the last ice age maximum ($\sim$17,000 years ago) from the present situation.

Today, the polar front is just south of Greenland (Fig. 7.2, $-2°$ C isotherm). During the glacial maximum, the front ran from New York to the Iberian peninsula. Norway, and even England, were entirely cut off from the warming influence of the Gulf Stream, with dramatic climatologic consequences.

The motion of the polar front in the North Atlantic, as reflected in deep sea cores, is a prime example for climatic transgression. Such a transgression can be used to determine the *rate* of change. How fast does a glacial period change into an interglacial one? How fast can the glaciers build up at the end of a warm period? These are questions of considerable interest, because mankind is involved in *climate modification*, by industrial CO_2 emission and by deforestation, and because the present unusually warm period (since 10,000 years ago) has been about as long as comparable ones in the past, throughout the Late Pleistocene. We shall return to these questions in Chapter 9 when discussing the geologic record.

7.2.4 Limitations of the Transfer Method. The method outlined (or others like it) allows an estimate to be made of any environmental variable which shows a correlation with foraminiferal abundances (or abundances of other plankton: coccoliths, radiolarians, diatoms). However, the significance of such estimates is in doubt where the environmental variable is not in control of growth. A prime example is salinity. There is little evidence that salinity has any influence on plankton distribution within the normal range of seawater salinity.

There are some other problems with the transfer method. First, exact calibration is difficult. Sediment arriving on the sea floor becomes thoroughly mixed with older sediment. Thus, the calibration set of sea floor samples contains information spanning thousands of years. Yet, the calibration set of present sea

163

surface conditions is based on a very few years at most. Are these last few years —
the ones whose data enter the present-day temperature atlas, for example —
representative for the last two or three thousand years? Perhaps.

Second, there is the problem of *differential dissolution*, hence *selective
preservation* of plankton shells on the sea floor. Most of the calcareous ooze on
the deep sea floor is exposed to at least some dissolution. Thus, the assemblages
from which we wish to draw paleoclimatic information are being altered on the
sea floor. The more delicate shells are being removed, while the thick-walled
forms are being concentrated (Fig. 7.6). This is also true for siliceous shells,
although dissolution patterns are different.

For older deposits the question arises: to what degree has evolution changed
the optima and ranges of the species? The question is ever-present in
paleoecologic research and is always difficult to deal with.

Benthic organisms, naturally, have the same basic requirements as planktonic
organisms — the right temperatures for growth and reproduction, sufficient food,
and other factors. Thus, climatic information enters benthic assemblages in much
the same way as planktonic ones, and can be similarly extracted, using the transfer
methods. However, while the remains of plankton on the sea floor are arranged in
broad climatic zones, benthic assemblages have a strong regional component. We
refer here to shelf assemblages, as the ones being most affected by climate.

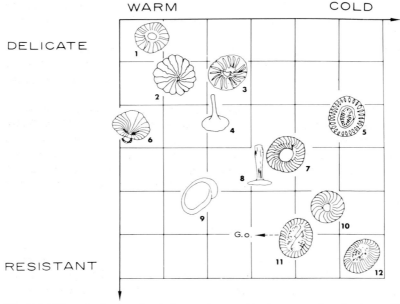

Fig. 7.6. Effect of selective dissolution on the temperature aspect of a mixed coccolith assemblage.
Warm water forms are removed and cold water species are enriched in the assemblage as dissolution
proceeds. *1 Cyclolithella annula; 2 Cyclococcolithina fragilis; 3 Umbellosphaera tenuis; 4 Discospha-
era tubifera; 5 Emiliania huxleyi; 6 Umbellosphaera irregularis; 7 Umbilicosphaera mirabilis;
8 Rhabdosphaera stylifera; 9 Helicopontosphaera kamptneri; 10 Cyclococcolithina leptopora; 11
Gephyrocapsa sp.* (includes *G. oceanica* and *G. caribbeanica;* the latter is shown); *12 Coccolithus
pelagicus.* (From W. H. Berger, 1973, Deep-Sea Res 20: 917)

7.3 Diversity and Shell Chemistry as Climatic Indicators

7.3.1 Diversity Gradients. One way to map changing climatic patterns is to focus not on the species themselves but on the statistical patterns they make: their *diversity* and their *dominance*. Diversity can be measured in various ways; the most realistic definition is the number of species in a standardized sample. For any sufficiently large group of shallow-water organisms, both benthic and planktonic, the number of species is high in the tropics, low toward the poles. There is, then, a diversity *gradient*, which runs roughly parallel to the global temperature gradient. In coastal areas, however, anomalously low salinity values or high influx of mud can adversely affect the marine habitat which can lower diversity independently of temperature.

Quite commonly, abundances of certain species are very high in areas of low diversity, because the physical rigor of the environment excludes competition or enemies or both. The great dominance of such species indicates environmental stress. Isolated lagoons, brackish marginal seas, and Arctic shelves illustrate the principle of low diversity and high individual abundance under conditions of physical stress. So do polluted lakes, for example.

7.3.2 Oxygen Isotopes. Another way to track climatic change in paleoclimatic reconstruction is to find relationships between shell chemistry and environment. The most widely applied method in this regard is the *oxygen isotope technique* of the great chemist H. C. Urey. Urey showed, some 30 years ago, that carbonates precipitated from the same aqueous solution should have different ratios of oxygen-18 to oxygen-16, depending on the temperature at which the precipitation proceeds.

The usefulness of this concept to paleoclimatic research was established soon after, by S. Epstein, R. Buchsbaum, H. A. Lowenstam, and H. C. Urey. They compared the isotopic composition of mollusc shells which grew at various temperatures, in their natural environment. Results showed that there was an orderly relationship between the isotopic composition and the temperature of growth. A fit to their data points yielded an equation which is widely used for paleo-temperature determination:

$$t = 16.5 - 4.3 \, (\delta_s - \delta_w) + 0.14 \, (\delta_s - \delta_w)^2 \tag{7.2}$$

The terms are as follows: t is the temperature, δ_s is the oxygen isotope composition of the shell sample, δ_w is that of the water the shell grew in. The δ notation describes the deviation of the ratio of oxygen-18 to oxygen-16 from that of a standard (for example mean ocean water), as a fraction of that of the standard:

$$\partial^{18}O = \frac{{}^{18}O/{}^{16}O \, (\text{sample}) - {}^{18}O/{}^{16}O \, (\text{standard})}{{}^{18}O/{}^{16}O \, (\text{standard})} \cdot 1000 \tag{7.3}$$

For convenience, it is expressed in "per mil", that is why the above fraction is multiplied by 1000.

Whenever an observed relationship between temperature and isotopic composition follows Eq. (7.2), the shell is said to have been grown in *isotopic equilibrium* with seawater. While molluscs (and planktonic foraminifera) generally precipitate shells in oxygen isotope equilibrium by this criterion, many organisms do not. Furthermore, the δ_w is variable geographically and through time, especially in coastal regions. It is closely related to the salinity of the ocean, because evaporation and precipitation affect both $^{18}O/^{16}O$ ratio and salinity patterns. Thus, unless *salinity effects* can be excluded, the paleotemperature cannot be deduced from the oxygen isotopes, other than within rather broad limits.

7.3.3 Other Chemical Markers. Besides oxygen isotopes, there are two stable *carbon isotopes* in calcareous shells: carbon-12 and carbon-13. These also have a temperature dependence, but the main effect on their ratio appears to be the metabolic activity of the shell-secreting organisms. Fast-growing organisms, and those harboring symbiotic unicellular algae (zooxanthellae), precipate shells with relatively low values of carbon-13.

Magnesium in calcareous shells tends to increase with temperature. This effect appears to be more pronounced in the lower organisms such as calcareous algae and foraminifera, than in the higher ones, for example the barnacles. The relationship has not been exploited for paleoclimate work to any significant degree.

Mineralogy follows temperature to some degree. There is a tendency for relatively more *aragonite* to be precipitated by benthic organisms in warm waters than in cold waters, where *calcite* is distinctly favored. Aragonite is the less stable form of calcium carbonate and changes to calcite during diagenesis. Thus, aragonite content of ancient calcareous sediments is largely a result of post-depositional alteration, rather than of conditions of formation.

7.4 Coral Reefs, Markers of Tropical Climate

7.4.1 Global Distribution. The classic geologic method of mapping climatic belts independently of individual species is to take typical shelf assemblages, such as *coral reef associations*. Even if the species within the associations have changed their preferences, as a whole, the reef association is believed to indicate tropical shallow waters. Today, the tropical ocean is the core region for coral reefs (Fig. 7.4). Reefs extend into the subtropics on the western sides of the ocean basins, and are greatly restricted on the eastern sides. The warm western boundary currents are responsible for carrying the reef limits to their latitudinal extremes, while eastern boundary currents bring cold water from high latitudes and are responsible for the restricted distribution of reefs on the eastern side of ocean basins. The reefs of the west-Atlantic island Bermuda reach to about 30° N,

166

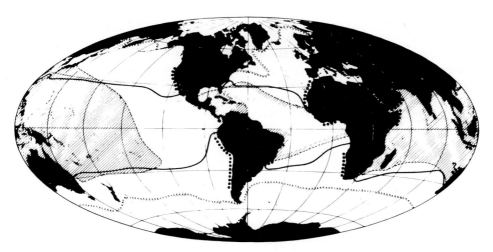

Fig. 7.7. Distribution of coral reefs *(hachured areas)*. 20° C isotherm for coldest month of the year shown as *heavy line*. Note the general correspondence. Areas of coastal upwelling *(dots)* inhibit coral growth. Hence the eastern margins are unfavorable for reef development. Present maximum Equator-ward extent of iceberg drift also is shown. (From E. Seibold in R. Brinkmann [ed], 1964, Lehrbuch der allgemeinen Geologie. F. Enke, Stuttgart; coral reef distribution after J. W. Wells, 1957, Geol Soc Am Mem 67: 609)

thanks to the Gulf Stream. On the eastern margin, in the sphere of influence of the Canary Current and of upwelling, reef organisms begin to grow only south of Dakar (15° N). However, on the whole the Equator does denote the mid-line of the distribution, an observation which is useful in paleogeographic reconstructions.

7.4.2 Reef Building. "Coral reefs" are built up by many different kinds of organisms and corals do not necessarily contribute the bulk of the material. Up to 10,000 grams of calcium carbonate per square meter can be produced annually, with a rate of upbuilding of 10 mm per year. Calcareous ooze, for comparison, builds up a thousand times more slowly, with accumulation rates near $10 \text{ g/m}^2/\text{yr}$. About 50% of the reef is solid matter and 50% is empty space (that is, pore space filled with water). Where the sea floor sinks, reefs can build enormous shelf bodies (on continental margins), or mountain peaks (on sunken volcanoes).

How is the reef built? To begin with, we have to distinguish between the *framework* (the supporting structure) and the *fill*, much as in building construction. The corals proper, and just as importantly the calcareous algae, are involved in producing the framework. Massive or (in protected places) branching stony corals can grow up by more than 2.5 cm/yr, as individual colonies. In the shallowest water, encrusting algae occur which grow over dead portions of the reef and cement it to such a degree that it can resist even the strongest surf. These encrusting algae, too, can reach a high rate of upward growth, over considerable areas. The fill between the framework consists of reef debris and the hard parts of

167

various other reef inhabitants. The fill can provide up to nine tenths of the reef mass.

The reef dwellers are legion, for the reef offers an extraordinary variety of life styles (Fig. 7.8). Solid substrate favors epibionts such as bryozoans and sessile foraminifera. Cavities offer shelter for fishes and crustaceans. The rich and colorful mollusc fauna is familiar from underwater photography and (rather unfortunately) from souvenir shops.

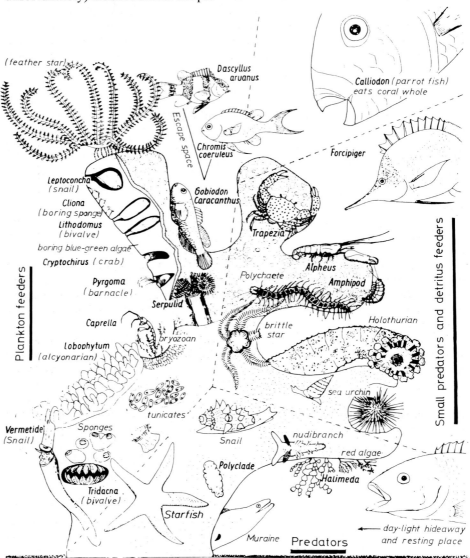

Fig. 7.8. Stone coral colony with associated macrofauna, classified as to mode of feeding. The dead basal part of the coral forms a hard substrate on which epibenthos settles. Note the many boring organisms which help destroy coral structures. (From S. Gerlach, 1959, Verh Dtsch Zool Ges 356)

168

The tropical coral reef is the shallow water habitat with the greatest number of species, that is, with the greatest diversity. More than 3,000 species live in Indo-Pacific reefs. The skeletal remains produced in rich variety are broken up in various ways, by boring algae, sponges and sea urchins. Much is crushed by fishes and crustaceans. In many skeletons, calcite crystals are held together by organic material. When it decays, such skeletons disintegrate to lime mud. Finally, waves and tidal currents grind up the relatively soft calcareous matter. The debris fills the interstices within the reef structure, collects in the lagoons behind the reefs, and washes up on the beach. Also, debris trickles down the sides of the reef into greater depths, building the reef flanks. Cementation of the debris starts very soon, as well as filling-in of pore space by cement. As observed on Bermuda and on Jamaica, cementation begins already 1 m below the reef surface.

7.4.3 Atolls. The various processes of reef building have their classic representation in atoll development. Palm-studded rings of coral and shell debris, enclosing a lagoon and sheltering it from the pounding surf, are quite common in the tropical South Pacific. Charles Darwin first explained their origin (Fig. 7.9). His theory was built on the idea of a sequence going from *fringing reef* to *barrier reef* to *atoll*. The fringing reef grows immediately next to the land. The barrier reef is separated from the land by a shallow lagoon. (A special case, of course, is the 2,000-km-long Great Barrier Reef lying 30 to 250 km off eastern Australia.) The atoll, finally, is an isolated quasi-circular ring of reefs, with an emergent portion of reef debris, the atoll island. In the center of the ring is a lagoon. Dimensions vary, a large atoll can be up to 40 km in diameter. Since the algae, both as carbonate producers and as symbionts of corals, need light, the coral reefs can only grow in very shallow water. An atoll, therefore, cannot have grown up from deep waters. This is the reasoning that led Darwin to propose a combination of slowly sinking sea floor and upbuilding of reef.

In the last 30 years, Darwin's theory was proven by drilling into several atolls. In the drillhole on Enewetak Atoll (Marshall Islands of the tropical Pacific) the basaltic basement, of Eocene age, was hit at about 1,400 m depth. Detailed investigation of the cores recovered from the Enewetak drill hole indicated that the reef had emerged from the sea on occasion: land snails, pollen, and spores from terrestrial plants are present. Also, there are signs of freshwater influence in the chemistry of the limestones, and there are hiatuses — missing sections indicating that erosion took place at times during the Tertiary. A similar situation is indicated on Bikini Atoll, 200 miles away. In both cases the hiatuses are at depths of around 200 to 300 m below today's reef surface, although on Enewetak there are also deeper ones.

Did the islands bob up and down as suggested by H. W. Menard? He proposed that as the sea floor moves over bumps or warm spots in the upper mantle, any islands on it will be raised on the upside of the bump, and will be dropped on the downside. Or must we assume that global sea level changes in Tertiary times produced the remarkable record within the atolls? To answer such questions we need to know more about the internal architecture of atolls in various areas.

169

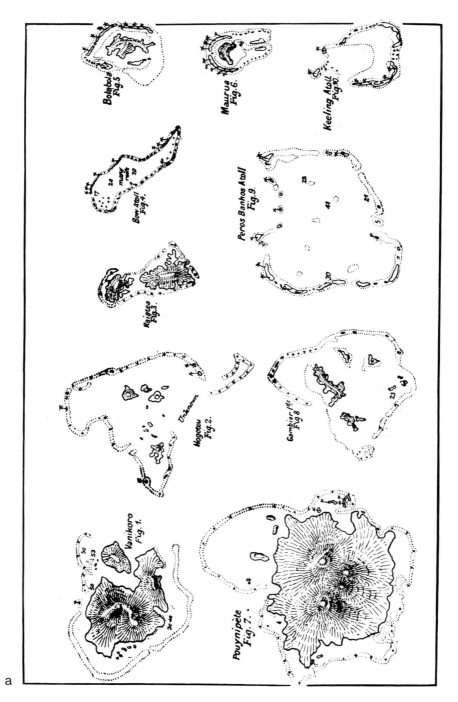

a

Fig. 7.9a. Darwin's hypothesis of atoll formation. **a** maps drawn by Darwin to show the resemblance in form between barrier coral reefs surrounding mountainous islands, and atolls or lagoon islands

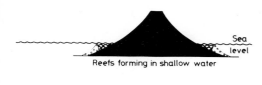

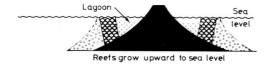

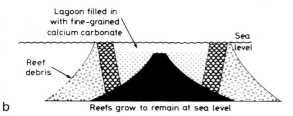

b

Fig. 7.9b. subsidence theory of atoll formation (After C. Darwin, 1842, On the structure and distribution of coral reefs, Ward Lock & Co, London; redrawn)

7.5 Geological Climate Indicators

7.5.1 Chemical Indicators. We discussed some of the aspects of biological indicators of climate. Coral reefs, of course, can equally well be considered geological indicators: they are calcareous sediment bodies whose patterns can be mapped. In fact, simple compilations of reef occurrences through time have been made from time to time to indicate the position of the equatorial belt and, ultimately, the cross-latitudinal motion of the continents. Biogenous carbonate deposits in general yield similar clues (Fig. 7.10). Although the method is not very reliable, it does provide some checks on geophysical results.

Salt and dolomite deposits also have been used to delineate climatic belts. They are characteristic for the subtropics and result from excess evaporation as discussed (Sect. 3.7.3 and 7.6). In tropical areas, laterite and the clay mineral kaolinite are typical weathering products. They find their way into deep sea deposits where they can be used to trace climatic changes on land (see Sect. 8.2.3).

7.5.2 Physical Geologic Indicators are especially obvious in high latitudes, due to the effects of ice (Fig. 7.10).

On shelves entirely covered by ice, as in the Antarctic, erosion predominates. Bare rock, polished and scratched, dominates; morainal debris fills depressions. Icebergs calving from glaciers take enclosed debris with them, and release it on

171

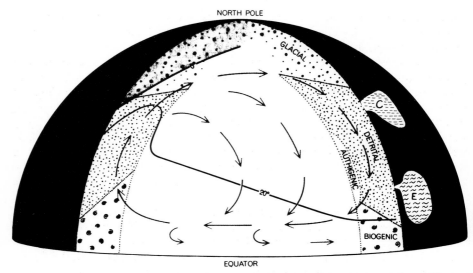

Fig. 7.10. Distribution of shelf deposits in relation to climatic zonation, in the northern half of an idealized ocean basin. Continents *black*. Currents shown as *arrows*. *Detrital* refers to terrigenous river influx, *authigenic* to phosphatic deposits associated with organic-rich sediments, *biogenic* to carbonates (molluscs, foraminifera, corals and algae). Shell carbonate is produced outside the tropics as well, but tends to be masked by terrigenous supply. Restricted shelf seas collect carbon-rich *(C)* or evaporitic sediment *(E)* depending on climate zone. (Modified from K. O. Emery, 1969, Sci Am *221* [3] 106)

melting. During the glacial periods of the Pleistocene, the area of the sea floor affected by iceberg sedimentation, was considerably expanded (Sect. 3.2.2). Also, extensive shelf areas were exposed at the time due to the sea level drop, and glacial ice could deposit end moraines in their typical lobe-shaped morphology, as on the East Coast north of New York and on the area of today's North Sea and Baltic. These moraines then are glacial relicts which are being worked over by shelf currents at the present time (Sect. 5.3).

Many high-latitude shelves (off Siberia, and off northern Alaska) are covered seasonally by sea ice. The rivers entering these shelf seas are commonly ice-covered, except for a few summer months. When the ice breaks up and the flood sets in, the freshwater may stay on top of the sea ice, off Alaska up to 10 km offshore. It penetrates downward through cracks in the ice and can dig 10 to 20-m-wide potholes on the sea floor below the places of entry.

The sea ice, 1 to 4 m thick, hinders the transport of sediment by dampening waves and longshore currents. The absence of sufficient longshore transport can perhaps explain the scarcity of lagoons in the Arctic. The sea ice also shields the entire Arctic sea floor from pelagic sediment supply: sedimentation rates in the central Arctic ocean are extraordinarily low ($\sim$ 1 mm/1000 years).

The ice can pile up, compressed by ocean currents, as so vividly described by Fridtjof Nansen (1861–1930), the great Arctic explorer. On the shelf the ice can then scrape the ground and produce deep, long ravines. These shelf areas are evidently distinctly unfavorable for the laying of pipelines or the building of platforms. Furrows produced by drift-ice during the last glacial have been discovered by side-looking sonar (Sect. 4.3.2) on the Scottish and Norwegian shelves out to the shelf rim.

Arctic environments are very demanding, as regards benthic life. Paucity of species is characteristic. One lagoon in Alaska southeast of Point Barrow, about 8 km wide and 4 m deep, is frozen over for 9 months in the year. On freezing, salt is largely excluded from the ice; thus the salinity of the remaining water is increased to more than 65‰. When the ice and the snow cover melts, in early summer, the water is diluted to a salinity of 2‰. When the sea ice melts, in mid-summer, normal marine waters enter the lagoon and salinity increases to 30‰. In October, the lagoon freezes over again, the cycle starts anew. Clearly, only a very few types of shell-producing organisms (or burrowers) can be expected under such extreme conditions.

The supply from rivers in high latitudes appears to consits mainly of silt, with clay and sand being distinctly subordinate. This is an interesting sedimentologic phenomenon, presumably it is caused by mechanical weathering through the freeze–thaw cycle. The process is important in producing *loess,* the silty sediment blown out from glaciated areas and piled up in their perimeter.

7.6 Climatic Clues from Restricted Seas

7.6.1 Salinity Distributions and Exchange Patterns. In contrasting tropical reef and ice-carved shelf, we have compared the warm and the cold extremes in the global ocean. There is another important contrast, that between regions of excess evaporation and of excess precipitation. These conditions are known on land as *arid* and *humid.* The ocean being water, the words "arid" and "humid" might seem to make little sense in describing the corresponding climatic belts in the ocean. However, we retain the terms as convenient descriptors of the balance between evaporation and precipitation.

Climate-produced salinity differences in the open ocean are present, and outline the major patterns of evaporation excess (central gyres) and precipitation excess (Equator, temperate to high latitudes) (see Fig. 7.11).

The salinity differences in the open sea are too small to leave much of a direct record via inorganic or biogenous sedimentation. However, the differences are amplified in the restricted seas adjacent to the ocean basins. We shall illustrate this amplification and its effects on sedimentation in some detail, believing that these concepts will be able to guide the interpretation of a large class of ancient marine sediments both from shallow and deep basins (see also Sect. 4.3.5).

The seas in the arid belt, with excess evaporation, have a common typical exchange pattern with the open ocean: shallow-in, deep-out (Fig. 7.12). This circulation pattern is called anti-estuarine. Prime examples are the Mediterranean, the Persian Gulf, and the Red Sea. Here, the loss of water by evaporation greatly exceeds the influx from rain and rivers. Thus, the sea level drops and water enters through a passage from the adjacent open ocean. The incoming water, of course, is derived from the surface water of the ocean, because it is at the surface

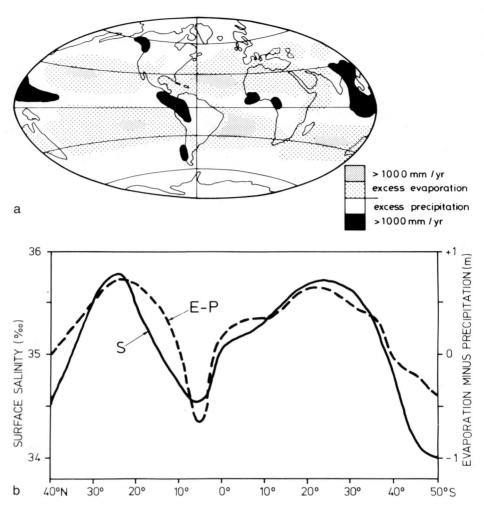

Fig. 7.11a, b. Evaporation/precipitation patterns in the world ocean, and corresponding salinity distributions. **a** Map showing difference between precipitation and evaporation (cm/yr) (E. Seibold, 1970, Geol Rundsch 60: 73, based on G. Dietrich, 1957). **b** Surface water salinities of the ocean, averaged by latitude, and compared with evaporation-minus-precipitation values. (After H. U. Sverdrup et al., 1942, The oceans. Prentice-Hall, Englewood Cliffs p. 124)

that the downhill gradient exists. The excess evaporation within the arid sea increases the salinity of the inflowing water which increases its density and makes it sink. Thus, the arid basin fills up with heavy, saline, surface-derived water. Being heavier than the open ocean water at the same elevation outside, the deep saline water pushes out and spills over the sill, at depth (Fig. 7.12). In this fashion the circulation of shallow-in, deep-out, is generated.

Where rain and river influx exceeds evaporation, sea level rises, and the water in the humid sea is freshened out and becomes *lighter* than the open ocean water.

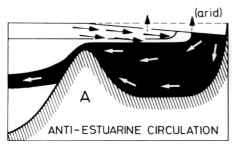
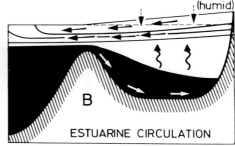

Atlantic – Gibraltar – Mediterr.
Indic – Bab el Mandeb – Red Sea
Indic – Hormus – Pers. Arab. Gulf

Atlantic-Sill-Norwegian and Greenland Fjords
Mediterr. – Bosporus – Black Sea
North Sea – Belts – Baltic

A ANTI-ESTUARINE CIRCULATION

B ESTUARINE CIRCULATION

Fig. 7.12. Anti-estuarine and estuarine circulation in basins with excess evaporation and with excess precipitation, respectively. The arid basin *(A)* is characterized by downwelling, hence low fertility and high oxygen content. The estuarine basin *(B)* is characterized by upwelling and salinity stratification, hence high fertility and low oxygen content. The geographic names above the graph give three examples each for anti-estuarine and estuarine circulation. (Modified after G. Dietrich, K. Kalle, 1957, Allgemeine Meereskunde. Borntraeger, Berlin)

Thus, the heavier ocean water pushes in at depth and displaces the less saline water from below. The incoming water is *subsurface* water. The outflow proceeds on the surface, where the gradient makes the water run down into the ocean. Examples for this type of circulation (deep-in, shallow-out) are the Black Sea, the Baltic, and the fjords of Alaska and Norway. This circulation pattern is called estuarine.

Whether a marginal basin has an "estuarine" or "anti-estuarine" circulation has profound consequences for its fertility and sedimentation. A comparison between the Baltic as the humid model and the Persian Gulf as the arid model will highlight the differences.

7.6.2 The Baltic as Humid Model of a Marginal Sea. First, the salinity distributions. In the Persian Gulf, surface *salinities increase* in going from the entrance near Hormus into the inner Gulf. The high salinity water *flows outward along the bottom.* In the Baltic, *salinity decreases* at the surface, from the entrance toward the inner bight. The heavy marine water *comes in at the bottom,* spilling from one depression to the next. In the Baltic low salinities are prevalent, generally less than 18%, due to dilution from rain and rivers. In the Persian Gulf salinities go up to 40%.

Both excessively low and excessively high salinities are unfavorable for normal marine organisms (Table 7.1). In the Baltic, the fully marine fauna off the entrance is gradually reduced in going deeper into the basin. Correspondingly, the dominance of certain brackish-tolerant species greatly increases. The marine molluscs become smaller and their shells thinner.

Another consequence of the humid condition is that the lagoons at the rim of the Baltic can easily freshen out, depending on precipitation, influx, and degree of

175

Table 7.1. Decrease of species from open ocean into Baltic (after Remane, 1958)

Marine Kinds of	Northsea	Baltic	Bay of Kiel	Middle Eastsea
Shells Marine	189	42	32	5
Snails	351	68	49	9
Cephalopoda	32	5	4	–

isolation. In such enclosed embayments, reeds and other plants grow to build up peat, eventually to be fossilized to coaly layers. The "Wealden" of England and NW Germany, a sedimentary sequence at the Jurassic-Cretaceous boundary, is a deposit of such an environment. The analogous boundary condition in the Persian Gulf, of course, are hypersaline lagoons and evaporite pans.

Finally, *stratification of the water column* is much more stable in the Baltic than in the Persian Gulf, due to the high density contrast between brackish surface water and saline deep water. In summer, when the surface waters in the Baltic are heated to 15° C or so, density difference and stability is further increased. Thus, vertical mixing virtually ceases. But without such mixing, no new oxygen can be supplied to the deep waters from the surface waters which are in contact with the atmosphere.

We can map the stratification using benthic foraminifera distributions (Fig. 7.13). Where the boundary between the more saline bottom waters and the more brackish surface waters touches the basin slope, a sharp boundary between upper and lower benthic fauna develops. A change in carbonate content and other sediment properties also occurs here — clues which can be useful for the interpretation of ancient marine sediments.

Productivity in the Baltic sea is high because nutrient-rich *subsurface waters* enter from the North Sea. The nutrients are trapped within the Baltic through organic sedimentation (calgae, fecal matter) and are partially recycled from the sea floor. They reach surface waters through mixing by winter storms, and by diffusion. The high supply of nutrients within the basin, however, leads to high productivity which in turn delivers much organic matter to the sea floor. Oxygen is rapidly used up by the decay of this matter. Especially during periods of stable stratification, the bottom water develops a serious oxygen deficiency, with values of less than 10% of saturation. Under certain conditions *all* the oxygen can be used up, and H_2S (the foul-smelling gas) develops from bacterial sulfate reduction.

7.6.3 Conditions of Stagnation. As the oxygen content drops, the bottom fauna changes. At a concentration of less than 1 ml O_2 per liter of water, shelled organisms disappear. Only a few species of metazoans remain: mainly worms (annelids, nematodes) and crustaceans. They stir the uppermost sediment, feeding on deposits, leaving burrows but essentially no hard parts. Below 0.1 ml O_2 per liter the environment becomes hostile to all metazoans; only some protists and anaerobic bacteria survive.

176

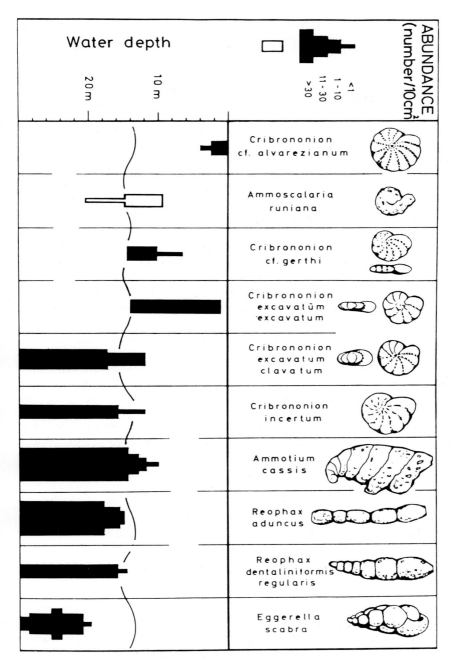

Fig. 7.13. Bathymetric distribution of the abundant benthic foraminifera in the western Baltic. *Width of black columns* indicates number of living specimens found per 10 cm² seabottom (*Ammoscalaria:* only empty shells were found in 1963/64). The *wavy line* at about 14 m depth marks the boundary between outflowing surface water and incoming marine water. (After G. F. Lutze, 1965, Meyniana 15: 75)

177

As mentioned, both high productivity and stable stratification are responsible for the oxygen deficiency in Baltic bottom waters. If nutrient supply is then involved in producing oxygen deficiency, could man's activities increase "stagnation" of the Baltic? Certainly. Deforestation and agriculture greatly increase soil erosion and hence the input of soil extracts, that is, nutrients. Also, modern agricultural methods rely on fertilizer which partly runs into the Baltic (and coastal waters in general). Sewage and industrial waste is yet another source. Thus, stagnation is indeed favored by man's activities. However, sediment cores from the central Baltic show that stagnation occurred frequently even before human factors could play a role. This is reflected in the portions of the record which consist of finely laminated sediments, with layers of millimeter dimensions (Sect. 3.9, Fig. 3.13). These layers indicate the absence of deposit feeders and other burrowers, hence they show the absence of oxygen.

The increase in stagnation in the central Baltic during the last 50 years, then, *could* be largely due to human activities, but it could also be a natural phenomenon which will eventually reverse itself. This example illustrates a very general problem in *environmental research and engineering,* namely that it is generally extremely difficult to keep apart the natural background fluctuations and the human influences.

The chemistry of anaerobic sediments is complex. Only a few aspects can be mentioned here. The CO_2 concentration in the bottom-near water with very low O_2 content is high: CO_2 is produced as the O_2 is used up. CO_2 and water produce carbonic acid, and the pH drops (to less than 7, compared with open ocean values near 8). Consequently, calcareous shells are dissolved on the sea floor. High CO_2 content of interstitial waters also means that hydrogenous carbonates can form. Manganese is readily mobilized under conditions of oxygen deficiency, migrates upward in the sediment as Mn^{2+}. It is kept on the sea floor both by oxidation and precipitation as oxide (MnO_2) and as *manganese carbonate* ($MnCO_3$) which forms under conditions of anaerobism and high CO_2 concentrations. Iron carbonate also may form, although iron is less mobile, being readily precipitated both as sulfide and as oxide. When free oxygen is gone, bacteria produce sulfide by the reduction of sulfate ion

$$SO_4^= + 2\,CH_2O \longrightarrow H_2S + 2\,HCO_3^-. \tag{7.4}$$

Iron sulfide forms under these conditions; hence pyrite crystals (FeS_2) are common in black anaerobic sediments. If sulfate reduction occurs at the very surface of the sediment, calcareous shells are preserved.

7.6.4 The Persian Gulf as Arid Model of a Marginal Sea. None of the conditions typical for the Baltic Sea develop in the Persian Gulf, because oxygen is brought to the sea floor by sinking saline waters, and because fertility is generally low. The *incoming ocean surface waters* are low in nutrients, as are warm surface waters everywhere. Hence the Persian Gulf is nutrient-starved. One place where anaerobic conditions can develop is in hypersaline lagoons, within the sediment. Sulfate reduction then occurs and organic-rich layers with pyrite can develop. However, such layers should be readily distinguishable in the geologic record from the stagnant basin layers in humid conditions.

178

The sediments of the Gulf proper are very different from those of the Baltic: organic contents (0.5% to 1%) are about five times lower, and carbonate content (>50%) is more than ten times higher. Benthic organisms exist at all depths and at all times, any lamination is quickly destroyed by bioturbation. Mobilization of heavy metals stops within the sediment because of the high oxygen supply in the mixed layer.

In the Persian Gulf the incoming surface water brings plankton from the Indian Ocean; some of it survives as the salinity increases within the Gulf; the remainder dies off. Planktonic foraminifera gradually diminish toward the inner Gulf, as seen in the ratio planktonic foraminifera/total foraminifera (Fig. 7.14).

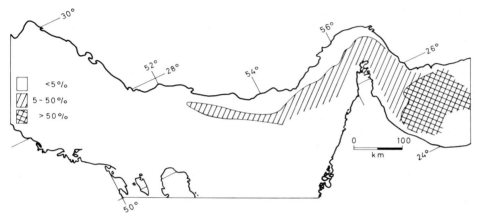

Fig. 7.14. Distribution of planktonic foraminifera abundance in the sand fraction of surface sediments of the Persian Gulf. The number shown is the percentage of planktonic foraminifera, of all foraminifera. The pattern indicates influx of plankton with surface waters entering from the open ocean *(at right)* into the Gulf. (Data M. Sarnthein, 1971, Meteor Forschungsergeb Reihe C 5: 1)

7.6.5 Application to Large Ocean Basins. The contrast between estuarine (humid) and anti-estuarine (arid) circulation is apparent even on the scale of very large basins (Sect. 4.3.5). The Mediterranean is the obvious example. Right now it is anti-estuarine, or arid, with sediments high in carbonate, low in organic matter and heavy metals. During early holocene black layers (sapropel) developed in the eastern Mediterranean. The most reasonable explanation is a freshening out of surface waters by runoff, and a reversal of the circulation — first proposed almost 30 years ago by B. Kullenberg, the Swedish oceanographer who developed the piston corer (Sect. 9.1). His suggestion has recently received much support from oxygen isotope studies indicating lowered salinities in surface waters, during sapropel periods.

On an even larger scale, the North Atlantic Basin is an anti-estuarine or arid basin, with sinking surface water, well-oxygenated bottom water, high carbonate and low organic matter content in sediments. Also, sediments have low opal content, as in the Mediterranean and in the Persian Gulf: diatoms and

179

radiolarians are readily dissolved. In contrast, the entire North Pacific can be seen as an estuarine-type basin, because of the low salinities reached in its northern belt (Fig. 7.15). Its sediments, in keeping with the Baltic model, are low in carbonate, high in diatoms and organic matter (at the slopes especially), and much of its deep ocean floor is covered by heavy metal concretions, the manganese nodules (Chap. 10).

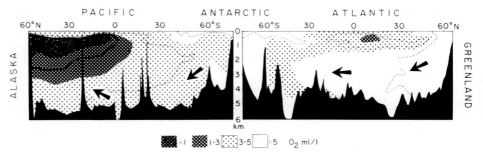

Fig. 7.15. Estuarine *(left)* and anti-estuarine *(right)* aspects of the deep circulation in Pacific and Atlantic ocean basins. The age of the deep waters is reflected in the distribution of the dissolved oxygen concentration. (From W. H. Berger, 1970, Geol Soc Am Bull 81: 1385)

8 Patterns of Deep Sea Sedimentation

Background. Deep sea deposits were first explored in a comprehensive fashion during the British Challenger Expedition a little over 100 years ago. Many thousands of samples were subsequently studied by John Murray (1841–1914), naturalist on the Challenger. He and his co-worker A. F. Renard published a thick volume on the results, which laid the foundation for all later work in this field of research. The first great step beyond Murray's work were the results of the German Meteor Expedition, almost half a century later. A new branch of oceanography started with the recovery of long cores by the Swedish Albatross Expedition, 1947–1949, that is, Pleistocene oceanography. It revolutionized our understanding of the great Ice Ages. Another great step came in 1968 with Glomar Challenger and the Deep Sea Drilling Project, which provided the samples for Tertiary and even Cretaceous paleoceanography.

We shall treat ancient ocean sediments separately (Chap. 9). Here we summarize the present-day patterns, as they appear in surface samples (Fig. 8.1).

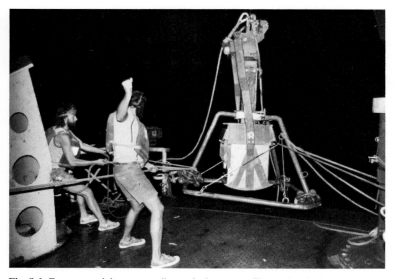

Fig. 8.1. Recovery of deep sea sediment by box corer. The coring device is a steel box with sharp edges which is pressed into the sea floor by heavy weights on top. The box is closed by a shovel which rotates around two bolts just above the box. The shovel is pushed downward when the box is pulled out, by pulling on the arm opposite the shovel. The frame with the three legs steadies the corer before it penetrates the sediment. The many lines visible in the photo prevent the heavy device from swinging on deck. (Photo T. Walsh, S.I.O.)

8.1 Inventory and Overview

8.1.1 Sediment Types and Patterns. First, an inventory of the types of sediment which are present (Table 8.1). The main types are already familiar from the last chapter. *Pelagic clays* are extremely fine-grained lithogenous and volcanogenic deposits. *Oozes* consist of biogenous minerals: shells of planktonic foraminifera, radiolarians, coccolithophores, and diatoms. *Hemipelagic deposits* are the same as clays and oozes except with large admixtures of shelf-derived sediment and of continental material. The list in Table 8.1 is not exhaustive, but includes the bulk of types found on the sea floor.

Table 8.1. Classification of deep sea sediments [from W. H. Berger (1974) in C. A. Burk and C. L. Drake. The geology of continental margins. Springer Heidelberg Berlin New York]

I. (Eu-) pelagic deposits (oozes and clays)
   <25% of fraction >5 μm is of terrigenic, volcanogenic, and/or neritic origin.
   Median grain size 5<μm (excepting authigenic minerals and pelagic organisms).
   A. Pelagic clays. $CaCO_3$ and siliceous fossils <30%.
      1. $CaCO_3$ 1–10%. (Slightly) calcareous clay.
      2. $CaCO_3$ 10–30%. Very calcareous (or marl) clay.
      3. Siliceous fossils 1–10%. (Slightly) siliceous clay.
      4. Siliceous fossils 10–30%. Very siliceous clay.
   B. Oozes. $CaCO_3$ or siliceous fossils >30%.
      1. $CaCO_3$ >30%. < $^2/_3$ $CaCo_3$: marl ooze. > $^2/_3$ $CaCO_3$: chalk ooze.
      2. $CaCO_3$<30%. >30% siliceous fossils: diatom or radiolarian ooze.
II. Hemipelagic deposits (muds)
   >25% of fraction >5 μm is of terrigenic, volcanogenic, and/or neritic origin.
   Median grain size >5 μm (except in authigenic minerals and pelagic organisms).
   A. Calcareous muds. $CaCO_3$ >30%.
      1.< $^2/_3$ $CaCO_3$: marl mud. > $^2/_3$ $CaCO_3$: chalk mud.
      2. Skeletal $CaCO_3$ >30%: foram~, nanno~, coquina~.
   B. Terrigenous muds. $CaCO_3$ <30%. Quartz, feldspar, mica dominant.
      Prefixes: quartzose, arkosic, micaceous.
   C. Volcanogenic muds. $CaCO_3$ <30%. Ash, palagonite, etc., dominant.
III. Pelagic and/or hemipelagic deposits
      1. Dolomite-sapropelite cycles.
      2. Black (carbonaceous) clay and mud: sapropelites.
      3. Silicified claystones and mudstones: chert.
      4. Limestone.

The broad outlines of deep sea sediment patterns are rather simple (Fig. 8.2). The major facies boundary in deep sea deposits is the calcite compensation depth, that is, the boundary between calcareous and noncalcareous sediments. Essentially, the calcareous facies characterizes the oceanic rises and elevated platforms, while the Red Clay facies is typical for the deep basins. Thus the overall pattern is depth-controlled. Superimposed on this pattern are the siliceous deposits, which accumulate below areas of high fertility; that is, the oceanic margins, the equatorial belt and the polar front regions. In addition, relatively

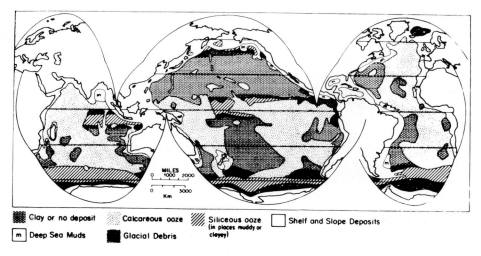

Clay or no deposit   Calcareous ooze   Siliceous ooze (in places muddy or clayey)   Shelf and Slope Deposits

m Deep Sea Muds   Glacial Debris

Fig. 8.2. Sediment cover of the deep-sea floor. The chief sediment types or facies are pelagic clay and calcareous ooze. (From W. H. Berger, 1974, in C. A. Burk, C. L. Drake [eds], 1974, The geology of continental margins. Springer, Berlin Heidelberg New York)

coarse terrigenous sediments from the continents invade the deep sea along the margins, in places to a considerable distance from the shelf, far into abyssal plains.

8.1.2 Biogenous Sediments Dominate. The bulk of the deep sea deposits consists of biogenous sediments, notably plankton shells (Table 8.1, Fig. 8.3). About one half of the deep sea floor is covered by *oozes,* that is, sediment formed from plankton remains. The plankton shells are generally less than 1 mm in diameter. The organisms producing the shells drift passively with ocean currents. Some migrate up or down, thus catching different horizontal currents at different depths. Except for some of the radiolarians, virtually all plankton organisms making sediment live in surface waters. Coccolithophores and diatoms need light for photosynthesis. This is also true for many planktonic foraminifera, because of symbiotic algae. Besides, food supply is highest in surface waters.

The shells and skeletons of foraminifera and radiolarians probably have several functions, including protection and trapping of food (on sticky protoplasm spreading along spines). One problem which arises for a plankton organism making mineral hard parts, is that these shells are much heavier than water, and tend to pull the organism down, away from sunlight and food. To counter this tendency, some skeletons are highly perforate, or delicate. Lipids and gas inclusions may increase buoyancy. Sinking tendency also is decreased by small size, and by the growing of spines and other protuberances.

Most shells of perished organisms never reach the sea floor, and most of those that do reach the floor are destroyed by dissolution. This is true both for calcareous and siliceous material, but especially for the latter. Virtually none of the ubiquitous open ocean diatoms are seen in sediments: they are too delicate to be preserved.

183

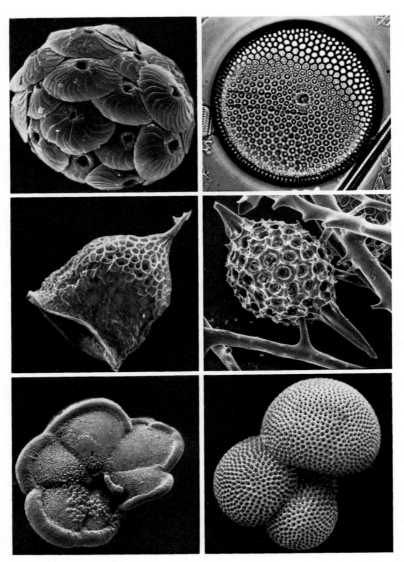

Fig. 8.3. Shell-bearing planktonic organisms. *Upper:* calcareous and siliceous algae; *left:* coccolitho-phore (×2600), with interlocking calcite platelets ("coccoliths"); *right:* diatom (×750), centric form from warm temperate waters. *Middle:* noncalcareous protozoans; *left* tintinnid (×600) with organic test; *right* radiolarian (×230), with siliceous skeleton. *Lower:* calcareous foraminifera; *left: Globorotalia menardii* (×35); *right: Globigerinoides sacculifer* (×65). (SEMPhotos by C. Samtleben and U. Pflaumann. Microphotograph *upper right* by H.-J. Schrader)

In summary, the most important factors controlling the composition of biogenous deep sea sediments are *fertility* and *depth*. Fertility, of course, controls, the supply of plankton remains, while depth controls the dissolution of carbonate (through pressure and water mass chemistry).

184

8.1.3 Sedimentation Rates. How fast do deep sea sediments accumulate? The first widely used rate estimates were those by W. Schott (1935). He identified the thickness of the post-glacial sediment in the deep Atlantic by noting the presence of *G. menardii,* a tropical foraminifera which is absent from the sections of glacial age (Fig. 8.3, lower left). Thickness divided by the age of the Holocene yields sedimentation rates. His method is still useful today, especially for shipboard determinations. Published rates (Table 8.2), show maximum values for terrigenous muds off river mouths (several meters per thousand years), intermediate ones for calcareous ooze (several centimeters per thousand years), and very low ones for Red Clay (a few millimeters per thousand years). These rates, ultimately, are based on dating of sediments through the measurement of radioactive isotopes within it (see Appendix A8) and through the correlation with paleoclimatic and paleomagnetic signals (see Sect. 9.3.5).

8.1.4 Thickness of Deep Sea Sediments. An important step in the investigation of the sea floor was the assessment of the total sediment thicknesses, by seismic (acoustic) refraction and reflection methods, in the 1950's. Early results of such surveys sent shock waves of surprise through the geologic community. The oceans were supposed to be a permanent, stable receptacle of continental and volcanic debris. M. Ewing and his collaborators at Lamont Geological Observatory found that the sedimentary columns for typical basins in the Atlantic are only about 500 m thick, and in the Pacific, a mere 300 m. It became clear that the old view of the deep ocean floor as a sedimentary memory reaching back for perhaps billions of

Table 8.2. Rates of accumulation of recent and sub-recent sediments [from W. H. Berger (1974) in C. A. Burk and C. L. Drake (eds.) The geology of continental margins Springer Heidelberg Berlin New York

Facies	Area	mm/1000 yr
Terrigenous mud	California Borderland	50–2,000
	Ceara Abyssal Plain	200
Calcareous ooze	North Atlantic (40–50° N)	35–60
	North Atlantic (5–20° N)	40–14
	Equatorial Atlantic	20–40
	Caribbean	∼28
	Equatorial Pacific	5–18
	Eastern Equatorial Pacific	∼30
	East Pacific Rise (0–20° S)	20–40
	East Pacific Rise (∼30° S)	3–10
	East Pacific Rise (40–50° S)	10–60
Siliceous ooze	Equatorial Pacific	2–5
	Antarctic (Indian Ocean)	2–10
	North and Equatorial Atlantic	2–7
Red clay	South Atlantic	2–3
	Northern North Pacific (muddy)	10–15
	Central North Pacific	1–2
	Tropical North Pacific	0–1

years had to be revised. In 1959, H. H. Hess could still entertain the idea that a hole from the sea floor to the Earth's mantle might sample primordial sediments on the down way. Shortly after, however (in 1960), he adopted the half-forgotten idea of mantle convection and constructed a model providing for a new sea floor "every 300 to 400 million years", which, he suggested, "accounts for the relatively thin veneer of sediments on the ocean floor" (see Chap. 1).

In the following we will take a closer look at the major types of deep sea deposits: Red Clay (in fact, reddish brown), calcareous ooze, siliceous ooze. Manganese nodules will be treated in the section on resources (Chap. 10).

8.2 Red Clay and Clay Minerals

8.2.1 Origin of Red Clay: the Questions. Of all deposits, Red Clay is uniquely restricted to the deep sea environment. The bulk of the components are extremely fine-grained, and the coarse silt and sand fractions consist of particles originating in the ocean: hydrogenous minerals, volcanogenic debris, ferromanganese concretions, and traces of biogenous particles such as fish teeth, arenaceous forams, and in some cases, spicules and radiolarians.

What is the source of Red Clay? This question is really three questions in one.

The first question addresses the ultimate source: to what extent is the clayey fraction in pelagic clays derived from in situ decomposition of volcanic material, and what is the contribution from continents and other sources?

The second question concerns transportation: what is the relative importance of transport by wind versus transport by rivers and ocean currents in bringing continent-derived clay particles to their site of deposition?

The third question considers chemical reactions between clay and seawater: to what extent are degraded clays from the continents "upgraded" by reactions with seawater, and what proportion of deep sea clays may be considered a "precipitate" from seawater? This third problem includes the concept of *reverse weathering,* which describes a process leading to the uptake of cations (Na^+, K^+) by clays (Sect. 3.3). The concept is implicit in the notion of an equilibrium ocean, as introduced by L. G. Sillén (1916–1970). Much of the postulated reverse weathering may take place in interstitial waters, especially within the more reactive hemipelagic deposits, perhaps involving intermediate biogenic precipitates. We will encounter the problem again when discussing siliceous deposits and the silica system.

8.2.2 Composition of Red Clay. To answer the first two questions (how much of the clay is of oceanic, how much of continental orgin; and how did it get there?) we need to know the composition of the fine-grained constituents, their distribution on the sea floor and their accumulation rates. Analyses by X-ray diffraction began in the 1930's (R. Revelle in the Pacific, C. W. Correns in the Atlantic) and has been greatly improved and systematically applied to deep sea deposits, especially in the last 20 years. The following minerals were found to be important in the clay and fine silt fractions of Red Clay:

186

1. Clay minerals: montmorillonite, illite, chlorite, kaolinite, and mixed-layer derivatives;
2. Lithogenous minerals: feldspar, pyroxene, quartz;
3. Hydrogenous (or authigenic) minerals: zeolite and ferromanganese oxides and hydroxides. The latter also provide much of the X-ray amorphous material present in Red Clay.

Concerning the lithogenous and hydrogenous minerals, they can be indentified also in the coarser fractions and compared to their parent rocks (e.g., basic or acidic volcanics, terrigenous rocks), so that their origin can be readily deduced. The distribution of quartz in the North Pacific suggests eolian transport from desert belts which agrees with evidence on size-distribution and chemical composition of the quartz.

The clay minerals warrant our special attention, since they make up the bulk of the finest size class ($\sim^2/_3$ of the clay size fraction) of nonbiogenous deep sea sediments, and the clay size fraction in turn provides some 90% of "pure" Red Clay. The main groups may be briefly introduced as follows (see Fig. 8.4). In the mineral group *montmorillonite* (or *smectite*), layers occur in a sequence such that an aluminous octrahedral layer is sandwiched between tetrahedral layers. Abundances of Mg^{2+} (also Fe) and Al^{3+} in the octahedral layer, and of Al^{3+} and Si^{4+} in the tetrahedral layer are such that there is a small net negative charge. This charge is balanced by exchangeable cations between the "sandwiches". The cations are hydrated, thus introducing variable amounts of water into the interlayer positions. This is the reason for the prime criterion for the identification of montmorillonite, that is, its ability to expand. Montmorillonite is a weathering product of volcanic rocks.

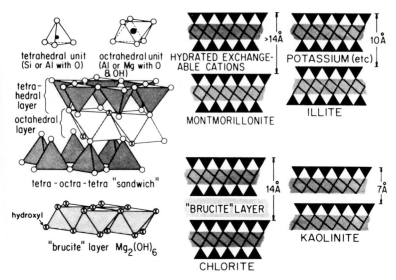

Fig. 8.4. Structure of clay minerals. The thickness of a complete layer is measured in Angstrom units ($\text{Å} = 10^{-8}$ cm), using X-ray diffraction techniques. Note the basic similarity of the clay minerals. (Mainly after R. E. Grim, 1968, Clay mineralogy. McGraw-Hill, New York)

187

Illite is a general term for clay components belonging to the mica group and their derivatives; for present purposes it may be considered a fine-grained degraded muscovite. The muscovite structure again shows an octahedral layer sandwiched between tetrahedral layers. There is only Al^{3+} in the octahedral layer, and the ratio of Si to Al in the tetrahedral layer is exactly 3 to 1. The net negative charge of the "sandwich" is balanced by nonhydrated, firmly bound K^+ ions fitting in the holes left by the hexagonal arrangement of the corners of the silica tetrahedrons.

Chlorite also consists of "sandwiches", but now bound by an additional octahedral layer, the so-called brucite layer. Net charges of tetrahedral layers and of octahedral layers balance each other, and there are therefore no inter-layer ions. Chlorite is a common constituent of low-grade metamorphic rocks which are widely exposed on glacially eroded shield areas and provide much of the material for glacial deposits.

Kaolinite consists of alternating tetrahedral and octahedral layers. It is a product of intense chemical weathering, and represents an insoluble alumino-silicate residue remaining after cations are stripped from feldspars and other minerals by extensive leaching.

8.2.3 Distribution of Clay Minerals. The clay minerals which are most abundant in deep sea clay are montmorillonite and illite (Fig. 8.5). Their distributions suggest that montmorillonite has important sources in oceanic volcanism, at least in the Pacific, while illite is largely derived from the continents. The remaining two important clay minerals, kaolinite and chlorite, also are land-derived, kaolinite from chemical weathering in the tropics and chlorite from physical

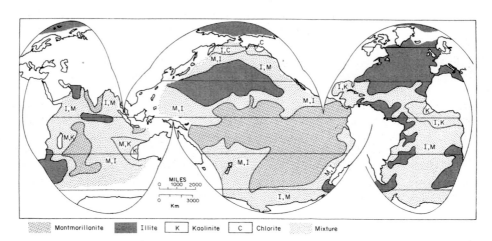

Fig. 8.5. Clay mineral distribution on the ocean floor. The map shows the dominant mineral in the fraction less than 2 µm. *Mixture* indicates that no one clay mineral exceeds 50% of the total. (From W. H. Berger, in C. A. Burk, C. L. Drake [eds], 1974, The geology of continental margins. Springer, Berlin Heidelberg New York. Main source: J. J. Griffin et al, 1968, Deep Sea Res 15: 433)

weathering in high latitudes. Off the west coasts of North Africa and of Australia, kaolinite actually becomes dominant. Chlorite is generally abundant off continents in high latitudes and is dominant in the Alaska Bight.

From the evidence reviewed, we may conclude that the case for continental sources of most clay minerals is very strong, and it remains only to ask what proportion of the montmorillonite is continent-derived. For the Atlantic, P. E. Biscaye (1965) argued that montmorillonite crystallinity patterns go parallel with detrital patterns and that this indicates a predominance of continental montmoril- lonite supply. In the Pacific, oceanic sources and "the ring of fire" supply the volcanic materials which decay to montmorillonite.

It is surprising that the clay mineral provinces are so clearly defined. Clay settles exceedingly slowly (see Sect. 4.1.2). If brought in by wind, therefore, it should have much opportunity to be dispersed by ocean currents. The very existence of the patterns calls for an explanation, therefore. M. N. Bramlette (in 1961) suggested that the removal of fine particles by filter-feeding planktonic organisms, and subsequent settling in fecal pellets, may contain the answer to such problems. Recent open ocean experiments on the transfer of fine particles to the sea floor show indeed that both biogenous matter and inorganic clay particles are quickly filtered out from surface waters by planktonic organisms, and incorporated into fecal pellets. In addition, near the ocean margins, bottom-near redistribution of clay minerals within hemipelagic sediment may play an important role in spreading the various clay types while retaining coherent patterns of distribution.

8.3 Calcareous Ooze

8.3.1 Depth Distribution. The rivers which feed the oceans are essentially dilute solutions of calcium bicarbonate and silica (Sect. 3.3). The input of calcium is such that the amount of calcium in the oceans can be delivered within about one million years. To balance this input, the ocean precipitates calcium carbonate. The precipitation takes place near the surface, within shell-building organisms (coccolithophores, foraminifera, molluscs, corals). Some of these shells (in the main, planktonic) find their way to the sea floor, where they are preserved on the more elevated parts and dissolved on the deeper ones, because the undersatura- tion of seawater increases with pressure and decreasing temperature (Fig. 8.6).

There are considerable differences in the carbonate distribution patterns of the Pacific and the Atlantic Ocean, with the Atlantic having higher carbonate percentages at all depths (Fig. 8.6b). Ultimately this difference is due to the effects of deep ocean circulation (Sect. 7.6.5; Fig. 7.15).

8.3.2 Dissolution Patterns in the Deep Sea. We would like to know at what rate the carbonate dissolves at each depth. Can we deduce the dissolution profiles for the Pacific and Atlantic from the carbonate patterns? Probably not, because a variety of different dissolution rate profiles can reproduce the commonly

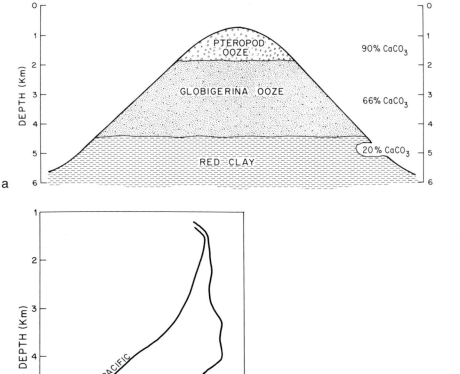

Fig. 8.6a, b. Depth distribution of calcareous deep sea sediments. **a** Idealized bathymetric zonation of deep sea deposits, produced by increasing dissolution of carbonate with depth. (according to J. Murray, J. Hjort, 1912, The depths of the ocean. Macmillan, New York). **b** Generalized depth profiles for carbonate content in deep sea sediments. (After R. R. Revelle, 1944, Carnegie Inst Wash Publ 556)

encountered carbonate-depth profiles. Thus the percent carbonate profiles based on core-top samples are poor indicators for dissolution rate profiles, especially in view of other factors such as variable dilution by clay-size noncarbonate material. We can use, however, the fragmentation of calcareous shells, and the selective removal of calcareous fossils through partial dissolution, as indications for the intensity of dissolution on the sea floor. This type of approach is promising, but it has not yet yielded maps for the rates of dissolution.

Most of what is known about global dissolution patterns on the sea floor is in fact represented by mapping the calcite compensation depth (CCD) (Fig. 8.7).

The *CCD* or *carbonate line* is analogous to the snowline which tends to follow a certain elevation contour in a given mountain range, at a given latitude. In concept, the CCD is the particular depth level at any one place in the ocean where

190

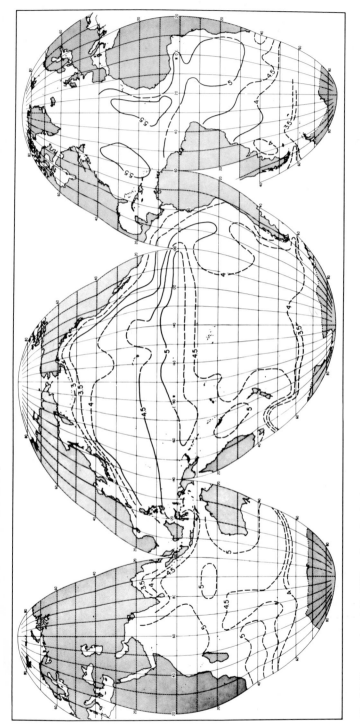

Fig. 8.7. Topography of the CCD surface, that is, the depth below which little or no carbonate accumulates. (From W. H. Berger, E. L. Winterer, 1974, Spec Publ Int Assoc Sedimentol 1: 11)

191

the rate of supply of calcium carbonate to the sea floor is balanced by the rate of dissolution, so that there is no net accumulation of carbonate (Bramlette, 1961). In practice, the CCD is mapped as the level at which percent carbonate values drop toward zero. This method can lead to difficulties in areas with pre-Recent carbonate exposures.

Another CCD-like level which can be mapped to describe dissolution patterns is the *lysocline*. The concept of the lysocline was introduced to denote a contour-following boundary zone between well-preserved and poorly preserved foraminiferal assemblages on the floor of the central Atlantic Ocean and on that of the South Pacific. In our snowline analogy, the lysocline corresponds to the boundary between fresh high-altitude snow and the wet or refrozen snow on the lower slopes.

Field experiments by M.N.A. Peterson showed a drastic increase of dissolution rates below 3500 m in the central Pacific (Fig. 8.8). The preservational lysocline on the sea floor correlates with Peterson's level in this area. In the Atlantic and in the South Pacific, the lysocline marks the top of the Antarctic Bottom Water. It appears very probable, therefore, that over large areas the lysocline denotes a level of increase in the agressiveness of the bottom water toward calcareous shells.

8.3.3 Dissolution Patterns Near Continents. Returning to the CCD map (Fig. 8.7), we note that its topography shows certain anomalies: the carbonate line does not simply follow depth contours. For example, high fertility along the Pacific equator leads to a depression of the CCD, by some 500 m. Paradoxically, high fertility *raises* the CCD in the margin areas around continents. The striking difference in content of organic matter between coastal and deep sea sediments offers a clue to this apparent contradiction. In the fertile areas of the ocean margins, the high supply of organic matter leads to highly increased benthic activity as well as to the development of much CO_2 in interstitial waters, producing carbonic acid. Thus, calcite shells are attacked even at depths of a few hundred meters on continental slopes. In the equatorial areas of the central Pacific, on the other hand, increased fertility leads to an increased supply of calcareous shells which goes well beyond the increased supply of organic matter in this region. The reason is that the organic matter produced in surface waters is also largely used there, that is, it is recycled rather than delivered to the sea floor. Whatever organic matter escapes this cycle and moves downward has a long way to go, and is largely used up within the water column. This leaves much less organic matter for sedimentation in the pelagic setting, than in the coastal one. Consequently, the ratio of calcitic shell to organic carbon produced is relatively high in the pelagic realm, which is favorable for the preservation of calcite.

8.3.4 Dissolution During Settling. The question of where calcareous shells dissolve, in the water or on the ocean floor, has been a matter of much speculation. Fecal pellet transport is important for coccoliths, which could not reach the sea floor if they had to settle through highly undersaturated waters as

192

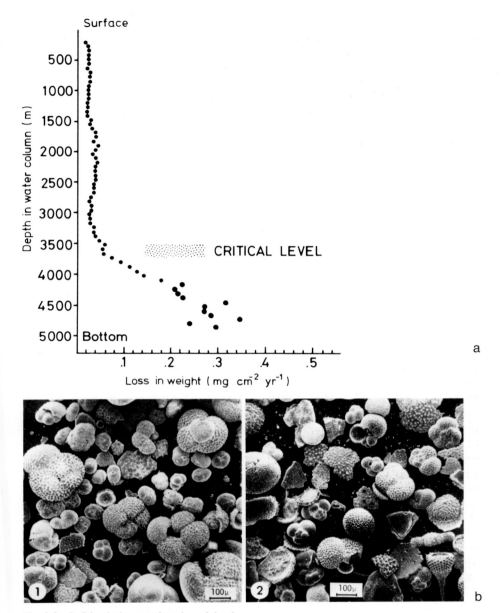

Fig. 8.8a, b. Dissolution as a function of depth.
a The Peterson Experiment. Polished calcite spheres were exposed on a line kept taut by a large buoy submerged below the surface. The line was anchored by a heavy weight. After 4 months the spheres were recovered and weighed. The diagram shows the weight loss (M. N. A. Peterson, 1966, Science 154: 1542).
b Selective dissolution. At the shallower depths on the sea floor, usually above 3,000 m or so, calcareous foraminifera are well-preserved (photo to *left*). As a critical level is approached, preservation deteriorates (photo to *right*). The boundary between well-preserved and poorly preserved foraminifera on the sea floor is the *lysocline*. It is closely associated with the critical level of Peterson. (SEMPhotos W. H. B./C. Samtleben)

193

single particles. Their settling rates are much too slow. For foraminifera, the evidence indicates that all but the small ones reach the sea floor, once they start to fall. This conclusion is mainly based on the study of box cores taken in the eastern Pacific, at depths well below the regional CCD. Examination of surficial sediments showed the presence of many delicate foraminifera, mixed with heavily corroded resistant ones and their fragments. Net tows at water depths below the level corresponding to the regional CCD level contained delicate forams, as well as the aragonitic shells of pteropods, which were not found in the sediment.

8.3.5 Controls on Carbonate Distribution Patterns. What factors ultimately control carbonate distribution? (1) We have seen that fertility distributions play an important role. Fertility regulates the supply of carbonate which, in the deep sea, ranges from some 0.6 g/cm^2/1000 years in the northern subtropical Atlantic to about 15 g/cm^2/1000 years in the eastern most tropical Pacific, with an average near 2 g/cm^2/1000 years. It also regulates the supply of organic matter which in turn largely determines the interstitial water acidity and the benthic activity, both inimical to calcite preservation. (2) Deep water circulation also is important. The Antarctic Bottom Water, either by reaching a *critical undersaturation,* or through rates of flow, appears to aggressively dissolve calcareous shells wherever present. The fact that abyssal waters of the North Atlantic are young (that is, have recently descended from the surface) and have had little time to accumulate CO_2 from the decay of organic matter, explains the unusually good preservation of deep sea carbonate in this region. The reverse is true for the North Pacific, which is filled with old, CO_2-rich deep waters (there are no sources for abyssal waters in the North Pacific at the present time). Hence, the CCD is deep in the North Atlantic, and shallow in the North Pacific.

The ultimate reason why abyssal waters dissolve calcite is that organisms supply calcium carbonate to the sea floor in excess of the amount that can be sedimented over the long run, an amount fixed by the influx from the continents and from hydrothermal sources. The shell supply to the ocean floor that exceeds this influx tends to deplete upper waters of calcium carbonate. Cooling of these upper waters and their subsequent compression and CO_2-enrichment at depth, during and after sinking, then produces bottom waters which are sufficiently undersaturated to redissolve the excess supply of calcium carbonate to the sea floor. Thus, a dynamic steady state is maintained. From this simple "book-keeping" concept, it can be readily inferred that, through geologic time, an overall increase in fertility leads to an overall increase in disolution, and vice versa.

8.3.6 A Global Experiment. At the present time, mankind is engaged in a global experiment involving carbonate dissolution, as well as climatic change. We — the industrial nations mainly — are burning off enormous amounts of coal and oil at an increasing rate. Large-scale deforestation proceeds in the tropics and elsewhere, for the sake of agricultural development and for wood products and fuel. The resulting carbon dioxide enters the atmosphere. So far, an amount equivalent to 20% of the CO_2 already present in the atmosphere has been added

194

within this century. One half of this added CO_2 has entered the ocean. Thus, it is estimated that the CO_2 content in the atmosphere is now 10% higher than at the end of the last century. Eventually, over the next few centuries, ten or twenty times the CO_2 in the atmosphere could be added to the air–ocean system (Fig. 8.9).

How will the ocean react to the input of CO_2?

In the long run, the ocean floor will neutralize most of the industrial CO_2 through the dissolution of carbonate:

$$CO_2 + H_2O + CaCO_3 \rightarrow Ca^{2+} + 2HCO_3^- \tag{8.1}$$

Thus, the industrial CO_2-pulse will produce a hiatus on the sea floor. A thickness of about 1 m of carbonate sediment will have to be dissolved from the carbonate-bearing sea floor, assuming that available coal and oil deposits are burned up. Initially, however, dissolution will mainly proceed in shallow areas, especially in high latitudes, where the water needs little additional CO_2 to become undersaturated. Subsequently it will affect low latitudes, presumably interfering with reef growth. It will take several hundred years for the deep areas of the sea floor to "feel" the effects of the increased CO_2 supply. It takes this long to renew the deep waters which bring the message of new atmospheric conditions to the deep sea floor.

8.3.7 A Prediction. Can we predict how high the CO_2 level will rise, and what the effects will be on climate? A great number of scientists — meteorologists, oceanographers, geochemists, and geologists — are working on this question. It is now generally agreed that a doubling of the CO_2 content would increase the global average temperature by about 2° C (4° Fahrenheit). Such a doubling might come within the next 50 years or so, depending on how the trends of use of carbon fuels develop. What might be the effect of such a temperature increase? Nobody knows the answer. It is possible that the climate would simply be warmer and

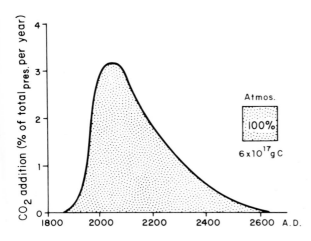

Fig. 8.9. One projection of input of CO_2 into the atmosphere, from the burning of fossil fuels (coal, oil). Annual input could reach 3% of present atmospheric content in the 21st century, if trends continue. The *shaded area* is the expected total input. (Based on M. K. Hubbert, 1969, in Natl Res Counc, Resources and man, W. H. Freeman, San Francisco)

195

moister over large areas in mid-latitudes, and that the change would be quite beneficial to agriculture. However, it is also possible that a marked warming will produce undesirable side-effects: a great increase in hurricane activity, for example, threatening densely populated coastal areas. A large-scale melting of Antarctic ice is conceivable, which could produce a great flood of biblical dimensions. Another possibility to be considered is that the (geologically) rapid CO_2 input introduces a strong disequilibrium into the ocean–atmosphere system, which provokes short-term climatic oscillations, something which is definitely inimical to economic and hence political stability.

The only safe prediction, from the point of view of geology, is that the CO_2 input will decrease again, probably within one to three centuries, either from exhaustion of resources, or from economic disruption caused by climatic change or other environmental problems (even neglecting the possibility of war).

The CO_2-problem may be an indication that we have come close to the limits of exploiting the natural cycles of our planet for man's use — yet a large part of a growing world population still lacks adequate food and shelter.

8.4 Siliceous Ooze

8.4.1 Composition and Distribution. In considering the siliceous deposits of the deep sea, many of the geochemical questions reappear which were raised earlier in connection with Red Clay and calcareous ooze. What are the contributions from continental weathering, submarine weathering, volcanic and hydrothermal emanations? What mechanisms control the concentrations of dissolved matter in seawater? That is, what controls the state of saturation? Are biogenous particles the sole sink for such dissolved matter, or is there uptake by "upgrading" of clays? What is the rate at which re-dissolution on the sea floor, if any, supplies material to the overlying waters?

First, let us take a brief look at the distribution, production, and dissolution patterns of the siliceous deposits.

We have already encountered the constituents of such deposits: remains of diatoms, silicoflagellates and radiolarians, and sponge spicules, all of which are made of opal, a hydrated form of amorphous silicon dioxide. Diatom oozes are typical for high latitudes, diatom muds for pericontinental regions, and radiolarian oozes for equatorial areas (Fig. 8.2 and 8.3). Both diatom and radiolarian ooze, of course, are mixtures of various kinds of sediments, with one or the other siliceous form being dominant (Fig. 8.10). The siliceous deposits typically occur in areas of high fertility; that is, in regions of surface water with relatively high phosphate values (Fig. 8.11). This overall correspondence between fertility patterns and silica-rich deposits can be considerably modified by redeposition processes within individual regions. The silica frustules are light and easily transported, and the activity of benthic animals which tends to resuspend fine sediment is especially pronounced in fertile areas. Thus, aided by bottom currents and gravity, siliceous frustules tend to accumulate in local and regional depressions.

196

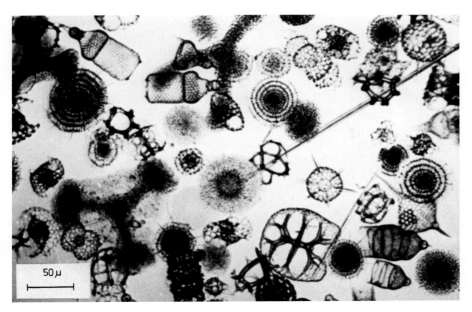

Fig. 8.10. Assemblage of modern radiolarians from sediments recovered in the equatorial Pacific. (Microphoto W. H. B.)

8.4.2 Controlling factors. In analogy to other kinds of deposits, the concentration of siliceous fossils in the sediment is a function of (1) the rate of production of siliceous organisms in the overlying waters, (2) the degree of dilution by terrigenous, volcanic, and calcareous particles, and (3) the extent of dissolution of the siliceous skeletons, most of which apparently occurs shortly after deposition.

The first variable, production of siliceous shells, attains its maximum in coastal regions (Sect. 4.3.3 also Fig. 7.3). This leads to the formation of a *silica ring* around each ocean basin. *Silica belts* are provided by the latitudinally arranged oceanic divergences which are a result of atmospheric circulation. The regions of divergence have nutrient-rich surface waters, hence there is sufficient silica available to make robust siliceous shells. Also, such areas are rich in grazing zooplankton, which pack the siliceous frustules into fecal pellets, thus accelerating delivery to the sea floor (Fig. 8.12).

To obtain an estimate of the amount of silica precipitated in the upper waters, one might multiply the measured amount of organic production with the ratio of solid silica to organic matter found in suspension. This yields only a rough estimate, of course.

A typical fixation rate of about 200 g SiO_2/m^2yr is suggested, with a range from less than 100 g (central gyres) to more than 500 g (Antarctic). Of this fixation, only 1 g/m^2yr, that is, 0.5%, can be incorporated into sediments if the river input is the only source of silica. Twice that (i.e., 1% of fixation) can be sedimented if we assume an equal contribution of silica from seawater–basalt reactions, especially at the hydrothermally active ridge-crests.

197

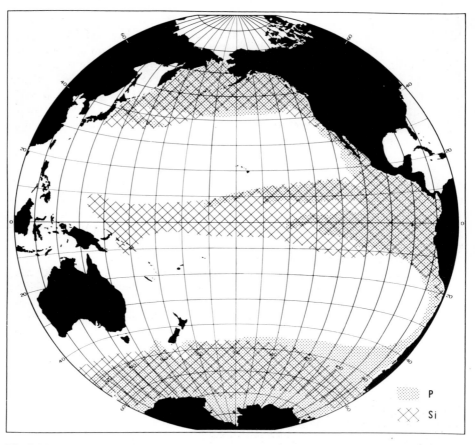

Fig. 8.11. Distribution of sediments rich in siliceous fossils *(Si)*, compared to distribution of fertile regions, where dissolved phosphate at 100 m depth is greater than 1 μg-atom per liter *(P)*. (From W. H. Berger, 1970, Geol Soc Am Bull 81: 1385)

The second factor, the degree of dilution of the siliceous material, reflects the ratio between accumulation rates of nonsiliceous and siliceous particles. As concerns dilution of silica by carbonate, one might expect that a high supply of calcareous shells would be accompanied by an equally high supply of siliceous shells, because both siliceous and calcareous plankton depend on productivity of upper waters. This is not generally the case, however. Indeed, there is a distinct negative correlation between silica and calcite distributional patterns. This has been ascribed to opposing chemical requirements for preservation. We have seen already that increasing fertility, from some point on, leads to decreasing preservation of calcite, but to increasing accumulation of silica. A similarly opposing trend is indicated for depth relationships, with silica corrosion being greatest in upper waters, that of carbonate being greatest at depth.

The third factor controlling the abundance of siliceous fossils is the extent of dissolution. The preservation of siliceous shells is rather closely correlated with

198

their relative abundance in sediments. A positive correlation between abundance and preservation may be ascribed to an increased supply of easily dissolved diatoms in fertile areas, which will "buffer" interstitial waters for the more robust skeletons, and to an otherwise favorable chemical environment in organic-rich sediments with slightly acidic interstitial waters. In general, silico-flagellates and diatoms tend to dissolve well before radiolarians and sponge spicules, and the following dissolution sequence can be established (from least to most resistant): (1) silicoflagellates, (2) diatoms, (3) delicate radiolarians, (4) robust radiolarians, (5) sponge spicules.

8.4.3 Geochemical Implications. The dissolution of opaline skeletons within the surficial sediment layer of the sea floor delivers silica to the deep ocean waters. This flux from the sediment to the water is evident from concentration gradients in interstitial waters and differences in concentration between these waters and the overlying sea water (Fig. 8.13).

It is possible and even probable that some of the silica released to the interstitial waters reacts within sediments to form new minerals. However, most of the silica reenters the ocean water. Thus, "old" bottom water, which has been in contact with the sea floor for a long time, is silica-rich. The reverse is true for "young" bottom water which has arrived from the surface only recently. Therefore, concentrations of dissolved silica are high in the deep North Pacific ("old" water), and are low in the deep North Atlantic ("young" water).

Fig. 8.12. Centric diatoms packed into membrane-bearing fecal pellet, by a copepod. Fecal pellet transport is important for transfer of diatoms, coccoliths, and clay particles to the sea floor. (Microphoto H.-J. Schrader, 1971; see Sci 174: 55, 1971)

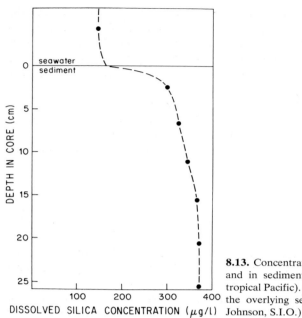

8.13. Concentration of dissolved silica in seawater and in sediment (box core taken in the eastern tropical Pacific). Silica diffuses from the sediment to the overlying seawater. (Measurements by T. C. Johnson, S.I.O.)

From this overall distribution of dissolved silica in deep ocean waters we can draw an obvious conclusion. The reason that silica concentrations in deep water are relatively low cannot be the uptake, if any, of dissolved silica by clay minerals. If it were, the "old" waters should be the more depleted in silica. The reason for the low concentrations must be that the deep water remembers its depleted condition at the surface (produced by silica extraction through diatoms) and that it did not have time to saturate itself with respect to the actively dissolving opaline shells.

8.4.4 Deep Sea Cherts. The discovery of chert within deep sea sediments has both fascinated geologists and frustrated them in efforts to drill and recover complete sections. The formation of deep sea chert (silicified sediments cemented by cryptocrystalline and microcrystalline quartz) appears to proceed from mobilization and reprecipitation of opal, generating a disordered cristobalite (=fibrous quartz) which eventually alters toward a quartzitic rock with mostly quartz-replaced and quartz-filled fossils as diagenesis progresses. Recrystallization may proceed at various rates and somewhat divergent patterns, depending on the original sediment present.

If siliceous fossils and/or silica-rich volcanic glass are to be available as a source for later production of chert, the following conditions appear necessary: (1) a sufficiently high supply of opal and low supply of dilutant; (2) silica-rich bottom water; (3) a reasonably high burial rate and (4) chemical conditions favorable for the preservation of siliceous shells.

200

During diagenesis, presumably, the opaline skeletons dissolve and volcanic material, if any, releases silica during devitrification, and the silica-rich interstitial solutions migrate along bedding planes or fractures and vertically to areas of precipitation in nearby permeable lenses or layers. Why precipitation should be favored in some places and mobilization in others in the same sediment is poorly understood in detail.

The question why deep sea chert deposits are concentrated in some geologic periods and not in others cannot be answered at this time. Presumably changes in the global silica supply to the ocean (weathering processes, volcanism, ridge-crest hydrothermal activity) changed the amount of silica deposited, and changes in ocean fertility changed the distribution patterns.

8.5 Turbidites

Turbidites are deposits generated by *turbidity currents* (Sect. 2.11 and 4.3.6). On the map of Fig. 8.2, these deposits are marked m. They are essentially terrigenous mud, spilled over from the continental margin into the deep sea.

The formation of turbidites depends on the availability of mud and a deep basin to put it in. In our time, the Quaternary, turbidites were generated especially during the ice ages.

During those periods the shelves were exposed and could not act as mud-traps for the rich sediment-load coming from land along much of the margin. Also, wave attack must have been strong. Storms and storm waves were probably more frequent than today, so that sediment on the outer shelf and on the upper slope

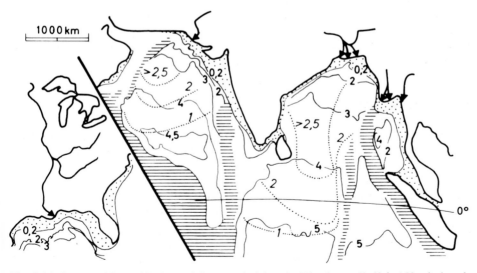

Fig. 8.14. Deep sea fans of Indus and Ganges, draining the Himalayas. Shelf (<150 m) *dotted;* *hachured areas* are ridges and rises in the deep sea. Isobaths (depth lines) in km. Isopachs (sediment thickness lines, *dotted*) in km. *Left* Mississippi drainage area and delta, same scale, for comparison

201

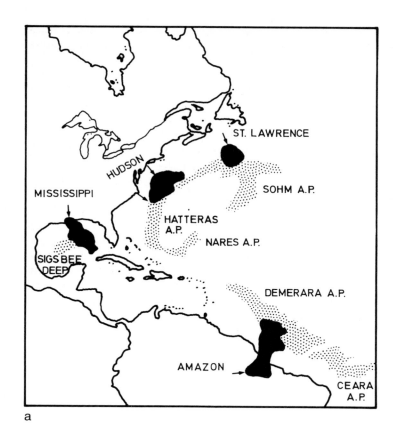

a

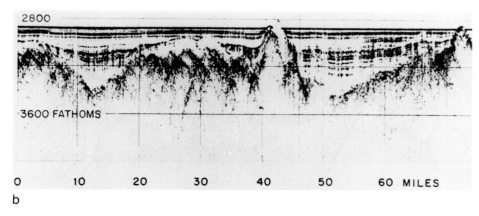

b

Fig. 8.15a, b. Abyssal plains. **a** Large deltas and deep sea fans of the Northwest Atlantic with associated abyssal plains. (After B. C. Heezen and M. Tharp, see Fig. 1.3). **b** Seismic echo profile across a stretch of abyssal plain. (Courtesy C. D. Hollister). Note that the sediment surface is perfectly horizontal regardless of the underlying basement topography. (2800 fathoms = 5100 m; 3600 fathoms = 6600 m)

202

could be re-suspended periodically, providing for mud-laden, heavy water bodies, which could then move down-slope, feeding submarine fans and abyssal plains.

The mud-laden turbidity currents cut into hard rock and into old sediments in the canyon part of a down–slope sea valley (Sect. 2.10). Here they erode rather than deposit sediment. Below the exit of the canyon a fan develops from dropped sediment and on the surface of the fan, distributory channels develop which are to 100 m deep. The channels have levees (see Figs. 2.12 and 4.1) and act much like a Roman aqueduct, funneling sediment into the deep sea, over very large distances (Fig. 8.14). The sediments form deep sea fans or large abyssal cones with widths of 200 to 2,000 km. Without the express-way mechanism, it is difficult to account for the immense distance of transport.

The very distribution of abyssal plains supports the idea that turbidity currents run along on the floor over extremely long distances (Fig. 8.15). For such currents, trenches would form an absolute obstacle, and we should not find abyssal plains beyond them. Once a trench is filled, however, plains can develop on the other side. This is exactly the distribution seen in the Gulf of Alaska. As the supply of turbidity deposits runs out, far in the deep sea, the abyssal ocean floor starts showing its typical bumpy morphology: the familiar abyssal hills, which are covered by pelagic oozes and clay.

9 Paleoceanography – The Deep Sea Record

In the chapter on climatic zonation, we have briefly referred to the record in several instances, especially in connection with the migration of the polar front in the North Atlantic (Sect. 7.2.3). In the last chapter, on deep sea sediments, we have seen how the ocean builds its memory. Let us now make a survey of what we have learned about the history of the ocean. The last fifteen years have witnessed an enormous amount of effort in this subject, especially through the work of the CLIMAP group in Pleistocene oceanography, and through the extensive drilling in the deep sea by Glomar Challenger (Fig. 9.1). In each case the studies involved the coordination and cooperation of marine geologists from

Fig. 9.1. The 120 m long deep sea drilling vessel Glomar Challenger, which has been working since 1968, sampling the sea floor of the world's ocean. More than 500 holes have been drilled, many in water depths of more than 5 km and commonly several hundred meters into the crust (occasionally penetration was more than 1000 m). The length of core material recovered measures tens of kilometers.
The Deep Sea Drilling Project, perhaps the most important earth science project of the 1970's, has been funded chiefly through the U.S. National Science Foundation; since the mid 1970's, the USSR, the Federal Republic of Germany, Japan, the United Kingdom, and France also have provided support. The project is managed by the Scripps Institution of Oceanography, with the advice and support of other major oceanographic institutions in the U.S.A. and abroad. (Photo courtesy Deep Sea Drilling Project, S.I.O.)

many institutions, in the USA and abroad. The harvest of these efforts is not yet wholly in, and new findings are constantly being reported. The following summary, therefore, is as much a survey of questions as of answers.

9.1 Pleistocene Oceanography

The idea that vast sheets of thick ice once covered northern Europe and northern North America became generally acceptable a little over one hundred years ago. Once this step was made, it was not long before *several* such glaciations were discovered, with warm intervals separating them. How many such advances and retreats of continental ice were there? What is the time-scale of these cycles? How do the cycles relate to astronomical factors, that is, the rotation of the Earth and its path around the Sun? These questions were difficult to even attempt to answer from the land record: each succeeding glaciation erased many of the traces of the previous one. The study of long cores from the deep sea floor opened new possibilites. Here there was a chance for a permanent record.

Long cores are recovered by dropping steel barrels into the sea floor; a weight on the upper end pushes them in. The barrel contains in its "nose" a piston, which remains stationary at the end of the wire, while the barrel falls past it into the sediment. With this arrangement, the sediment tends to stay put while the barrel pushes past, despite friction, because of the pressure deficit which develops below the piston, and which pulls on the sediment like a suction pump (Fig. 9.2).

The Swedish Albatross Expedition (1947–1948) made extensive use of this device (developed by B. Kullenberg) and recovered many long cores ($\sim$7 m) of calcareous ooze which contained a record of the last one half million to one million years. The study of these cores, by geochemical and micropaleontological techniques, yielded much information on the behavior of the ocean through the Pleistocene. Such studies have continued, on cores collected subsequently by many institutions, providing some answers to questions such as these:

1. What did the ocean look like during maximum glaciation?
2. What is the nature of the ice age cycles, that is, their frequency and amplitude, and their extent in time?
3. What can we say about the role of the ocean in the dynamics of change from interglacials to glacials, and vice versa?

We shall now take up these topics one by one.

9.2 The Ice Age Ocean

9.2.1 Conditions in a Cold Ocean. How did the ice age ocean differ from the present one? The question is by no means settled, but a consensus has been reached on several aspects.

First, it is generally agreed that *surface currents were stronger*. It is obvious why this should be so: surface currents are driven by winds, and winds depend on

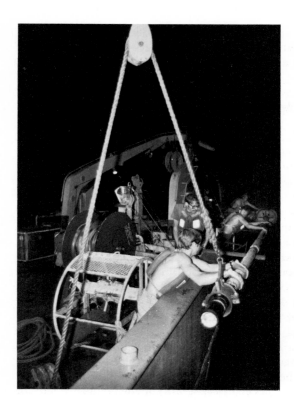

Fig. 9.2. Recovering the Pleistocene record by piston coring. The corer is a wide-diameter model; note the white sediment in the core nose. Note also that the core barrel was bent during the operation (on the upper end). Other equipment on deck: deep sea camera frame (with protective grid), hydrophone for seismic profilling (wrapped on spool), box corer (aft). (S.I.O. Eurydice Expedition, 1975)

horizontal temperature gradients. With the ice rim and the polar front much closer to the Equator, the temperature difference between ice ($0°$ C or less) and the tropics ($\sim 25°$ C) was compressed into a much shorter distance than now. Hence the temperature gradient was greater, winds were stronger, and so were ocean currents.

Equatorial upwelling was intensified as a consequence, and also *coastal upwelling*. Thus, at the same time that fertility decreased in high latitudes, due to ice cover, it increased in mid, latitudes (because of intensified mixing) and in the subtropics (due to upwelling).

Second, it is accepted that the *ocean surface was cooler,* on the whole, than today. With a substantial part of northern continents and seas under ice, the Earth reflected the Sun's radiation more readily (had a higher *albedo*) than today, hence, it absorbed less of the radiation and its atmosphere was cooler. A cold atmosphere holds less water than a warm one and large areas on land therefore were drier than today. Dry areas (such as grasslands and deserts) reflect more sunlight than wet ones (such as forests). Also, a more fertile ocean would be slightly more reflective than a clear dark blue one with less algal growth. All these factors favored reflection of the Sun's radiation back into space, and hence favored cooling.

206

Third, it is known that some ocean regions cooled more than others, and that the regional degree of change depends closely on the *movement of boundaries* of climatic zones. If, for example, an area is near the boundary of subtropics and temperate belt, it will then belong alternately to one or the other zone. Thus, here the changes are substantial. Conversely, changes can be minimal in the center of tropical or subtropical climatic regions.

9.2.2 The 18 K Map. Recently these various concepts were put to a test by a coherent quantitative reconstruction of the Ice Age ocean. A. McIntyre, T. C. Moore, and colleagues, using the "transfer" techniques of J. Imbrie (Sect. 7.2.2), produced a map of sea surface temperatures for a typical (northern) summer month, during maximum glaciation. The map is also called 18 K map, for "18 kilo years ago" (see Fig. 9.3). The construction of the map proceeded from deep sea cores, from which both surface and subsurface samples were taken. Surface samples were used for calibration (Sect. 7.2.1). The abundance patterns were correlated to surface water temperatures in northern summer, in this case.

The sampling depth in the cores for the 18 K age were found by various stratigraphic techniques, mainly oxygen isotope stratigraphy. In this method, the first sample, on going down-core, which shows an extremely oxygen-18 rich composition of calcareous shells is taken as the one representing the glacial (see Sect. 7.3.2). Each 18 K sample integrates a time interval which is between 400 and 4,000 years long depending on whether the sedimentation rate is high or low. After counting the species in these 18 K samples, the transfer equation (Sect. 7.2.2) can be applied, and thus the most probable temperature of the surface water at the time can be estimated.

From the information shown in the 18 K map, it can be concluded that the ice age ocean 18,000 years ago was characterized by: 1. increased thermal gradients along polar fronts, especially in the North Atlantic and Antarctic; 2. Equator-ward displacement of polar frontal systems; 3. general cooling of most surface waters, by about 2.3° C, on the average; 4. increased upwelling along the Equator in Pacific and Atlantic; 5. increased coastal upwelling and strengthening of eastern boundary currents; and 6. nearly stable positions and temperatures of the central gyres in the major ocean basins.

9.3 The Pleistocene Cycles

9.3.1 The Evidence. The marine geologists who first studied the long cores recovered by the Albatross Expedition (G. Arrhenius, F. L. Parker, F. B. Phleger) soon recognized that the Pleistocene record shows a long series of alternating climatic states. This finding has attained great importance in the earth sciences, especially in the study of climate dynamics. The cycles express themselves as fluctuations in faunal and floral composition, in the abundance of carbonate, and in the content of oxygen-18 of foraminiferal shells and in other properties (Fig. 9.4).

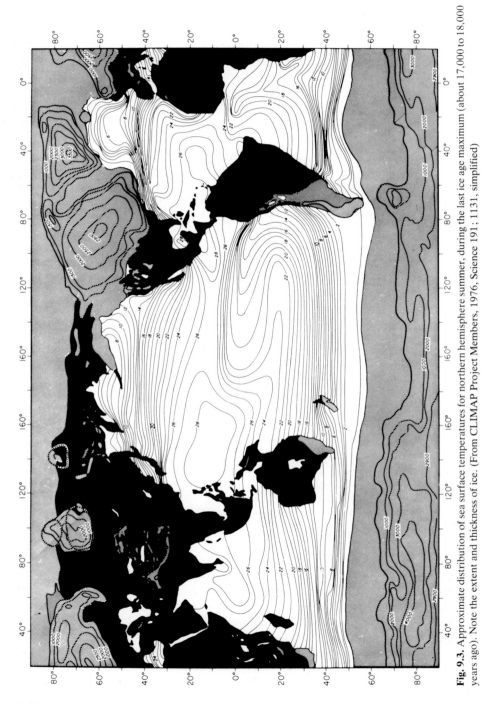

Fig. 9.3. Approximate distribution of sea surface temperatures for northern hemisphere summer, during the last ice age maximum (about 17,000 to 18,000 years ago). Note the extent and thickness of ice. (From CLIMAP Project Members, 1976, Science 191; 1131, simplified)

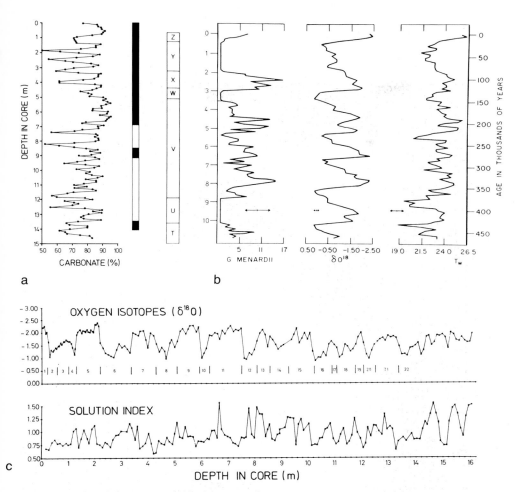

Fig. 9.4a–c. The Pleistocene record in deep sea sediments.
a Carbonate cycles of the eastern Pacific, magnetically dated (J. D. Hays et al., 1969, Geol Soc Am Bull 80: 1481). **b** *G. menardii* pulses *(left)* and transfer temperature cycles *(right)* compared with oxygen isotope stratigraphy, Caribbean Sea (J. Imbrie et al., 1973, J Quat Res 3: 10).
c Dissolution cycles compared with oxygen isotope stratigraphy, western Pacific (P. R. Thompson, 1976, J Foraminiferal Res 6: 208)

9.3.2 The Carbonate Cycles of the equatorial Pacific (discovered by G. Arrhenius) are referred to as *dissolution cycles,* with high dissolution in interglacials (low $CaCO_3$ values) and low dissolution in glacials (high $CaCO_3$ values). Productivity variations (which were once thought to be responsible for the carbonate fluctuation) play a secondary role in producing the cycles.

The causes for the dissolution cycles are not clear. Fluctuation of the sea level must be important. During low sea level, shelves — the preferred site for carbonate deposition — are exposed. There is no place for carbonate to go except

209

onto the deep sea floor. Conversely, during interglacials the high sea level allows shallow water carbonates to build up — this carbonate is extracted from the ocean, and is no longer available for deposition on the deep sea floor. Several other factors may play a role: changes in erosion rates on land, growth and decay of forests on land which affects the abundance of CO_2 in the ocean-atmosphere system, changes in the temperature of the deep ocean, changes in the overall fertility of the ocean, changes in deep sea circulation. These are matters of active scientific discussion today, and are far from being settled.

In the Atlantic the carbonate cycles swing counter to those in the central Pacific. The Atlantic carbonate cycles are largely *dilution cycles*. During glacials the supply of terrigenous materials from the continents surrounding the Atlantic is greatly increased. In high latitudes the activity of the ice grinds up enormous masses of rock. The plant cover, which protects soil from erosion, is less dense. In the subtropics deserts are wide-spread, delivering dust. Flash floods in semi-arid regions are efficient conveyors of huge amounts of material. The tropical rain forest is much reduced, and semi-arid regions are expanded. The shelves are exposed, and subject to erosion. All these factors contribute to the glacial increase of terrigenous deposition rates. As the supply of terrigenous material increases, of course, the proportion of carbonate in pelagic sediments decreases accordingly — the carbonate is diluted. By changing the degree of dilution, carbonate cycles are produced.

9.3.3 The Faunal (and Floral) Cycles can be expressed in various ways, most commonly as *warm–cold cycles*. Examples are the quantitative plots of Parker (1958) and of Imbrie and Kipp (1971) (Fig. 9.3b). Parker's method contrasts the relative abundance of warm-water and cold-water planktonic foraminifera, as a function of core depth. The Imbrie-Kipp method calibrates the warm- and cold-water percentages against surface water temperature, by a statistical technique called factor-regression, and then calculates the most probable temperature of surface water down-core. It is the same technique used for constructing the 18 K map (Fig. 9.2). We see that there were considerable temperature variations in the Caribbean. We also see that the last change from cold to warm occurred about 11,000 years ago. This is the end of the last glacial and the beginning of the present interglacial (the Holocene).

The faunal cycles have an important message, in that they are well defined in their amplitudes. Also there is an indication that they tend to avoid intermediate values, suggesting that there are two preferred states: warm, and cold, but not in-between. They show one other interesting fact. The present time is rather unusual in being so warm: for the last one half million years or so the climate was mostly much more severe.

9.3.4 Oxygen Isotope Cycles. The fluctuations in the oxygen isotope composition of foraminiferal shells were first described by C. Emiliani in his classic paper *Pleistocene temperatures,* published in 1955. Today, oxygen isotope stratigraphy with its "isotopic stages" forms the backbone of Pleistocene stratigraphy.

210

Emiliani analyzed the planktonic foraminifera from several long cores, taken in the Caribbean and North Atlantic. He concentrated on those species with the lowest oxygen-18 (*Globigerinoides ruber* and *Globigerinoides sacculifer*), reasoning that these species must live in shallow water and therefore reflect surface water temperature. Since the temperature of growth affects the $^{18}O/^{16}O$ ratio (Sect. 7.3.2), the isotopic fluctuations reflect warm–cold cycles. In addition, the composition of seawater controls $^{18}O/^{16}O$ ratios. The seawater composition fluctuates with the waxing and waning of the continental ice sheets. The reason is that ice is depleted in ^{18}O. During build-up of the ice, therefore, ^{18}O stays preferentially in the ocean, increasing the $^{18}O/^{16}O$ ratio. Shells grown in such water, therefore, are enriched in ^{18}O. (This enrichment with ^{18}O is added to the enrichment caused by the lowered temperature of the water.) When the glacial ice melts, the $^{18}O/^{16}O$ ratio in the ocean decreases again (see Fig. 5.15).

9.3.5 Milankovitch Cycles and Dating. As pointed out by Emiliani, the isotopic variations clearly indicate some sort of regular cycling such as could be produced by the *Milankovitch mechanism* which invokes regular variations in the Earth's orbital parameters as a cause for the succession of ice ages separated by warm periods (Fig. 9.5). The hypothesis of Milutin Milankovitch (1879–1958) states that long-term fluctuations in the radiation received from the sun, during summer seasons, in the high latitudes of the northern hemisphere, have controlled the occurrence of ice ages over the last 600,000 years.

Before Emiliani's suggestion could be tested, a time-scale for the isotopic variations was needed. Three dating methods are available: (1) C-14 dating of the uppermost portion of the record, and extrapolation downward. This method is not very reliable. (2) Uranium dating of corals which grew during the last high stand of sea level, a datum that can be correlated with the warm peak in the isotope stage 5. The best estimates for this age are near 124,000 years. From this we obtain an average sedimentation rate, which is more reliable than the one based on carbon-14. (3) Magnetic reversals — the same which are recorded in the cooling basalt of the spreading sea floor (Sect. 1.9) — are recorded also in deep sea sediments. The last major boundary between *magnetic epochs* (the *Brunhes* and the *Matuyama*), is at 700,000 years ago. This date can be recovered in very long cores only. It allows *interpolation* of the ages for the isotope variations.

The time scale derived from methods (2) and (3) is the one used in the comparison between isotope record and Milankovitch curve, in Fig. 9.5. A certain similarity of the curves is obvious. Spectral analysis — that is, a search for the frequencies contained in the isotopic record — shows that the following periods are strongly represented: 20,000 years, 40,000 years, and 100,000 years. These are also close to the main periods contained in the Milankovitch irradiation curve. Thus, the evidence is strong that irradiation of the northern hemisphere is the dominating factor in controlling the *frequencies* of the Pleistocene climatic fluctuations.

What are these frequencies? They describe the motions of the Earth's rotational axis and the change in shape of its path around the sun. The axis is not stationary in space, and does not always point to the North Star as at the present.

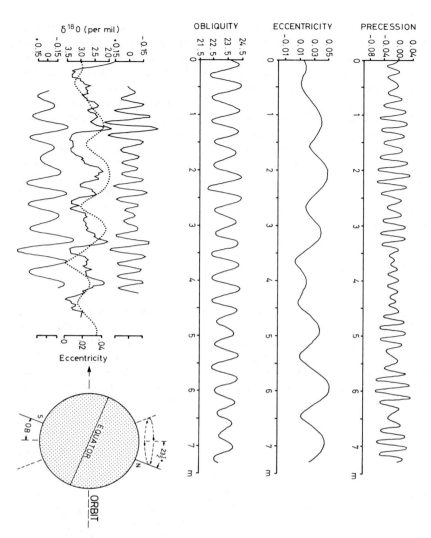

Fig. 9.5. The Croll-Milankovitch Theory of the Ice Ages, and its test. *Upper* 3 curves showing orbital parameters as calculated by A. L. Berger, in M. A. Kominz et al., 1979, Earth Planet Sci Lett 45 : 394. The eccentricity is the deviation of the orbit from a perfect circle, for which e = 0. The eccentricity varies with a period near 96,000 years. The precession parameter is a function of both the position of perihelion (point of Earth's orbit which is closest to the Sun) with respect to the equinoxes (day = night positions) and the eccentricity of the orbit (which makes the difference in positions of equinoxes and perihelion significant in terms of seasonal irradiation). The periodicity of the precession parameter is near 22,000 years. Note the small variations during times of low eccentricity. The obliquity is the angle of the Earth's axis, with respect to the vertical on the orbital plane (see *inset, lower right*). It varies with a period near 41,000 years.

Lower 3 curves showing the orbital periodicities extracted from the oxygen isotope record in a sub-antarctic deep sea core. (From J. D. Hays et al., 1976, Science 194: 1121). *Middle* Isotope record (solid) with superimposed eccentricity variation. *Top* 23,000 year component extracted from the isotope record by band pass filter (a statistical method), *Bottom* 40,000 year component extracted in a similar fashion. *Inset to right* illustrates obliquity

212

Instead, it describes a circle, of which the North Star is one point. The circle is completed once in about 22,000 years. This ist the *precession* (Fig. 9.5). Also, the inclination of the Earth's axis to the plane of its orbit changes through time. It is 66½° at present, but varies between about 65° and 68° once in 41,000 years. This is the obliquity variation. Obliquity is very important, because high obliquity obviously means warm summers and cold winters, and vice versa. Finally, the Earth's orbit about the sun is not a circle but an ellipse (as Johannes Kepler, 1571–1630, showed). The ratio between the long and the short axis varies through time. This is the "eccentricity" variation. One cycle takes about 100,000 years. These, then, are the elements governing the changes in the seasonal irradiation of the northern hemisphere. Presumably, there is an especially sensitive latitudinal belt in the northern hemisphere where these changes are translated into varying amounts of snow cover, which in turn govern climate by the albedo feedback mechanism (Sect. 9.2).

9.4 Dynamics of Change

Let us accept the idea that the irradiation cycles indeed correlate with the isotopic record, and that the Milankovitch mechanism is a reasonable way to explain the frequencies seen. We must then turn to two other important properties of the isotopic record: the stabilization of the maximum amplitude, and the "saw-tooth" shape of the curve (Fig. 9.6). Neither of these properties is at all understood, because the dynamics of atmosphere–ocean interaction cannot yet

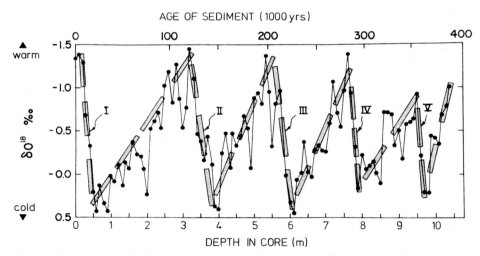

Fig. 9.6. The "saw-tooth" effect in the Pleistocene climatic record. The long-term variations of the oxygen-isotope record in deep sea cores (here: C. Emiliani, 1966, J. Geol 74: 109; Caribbean Sea) tends to show asymmetry, with short periods of very rapid warming (terminations, *I, II, III, IV, V*) and longer periods of general cooling. (From W. S. Broecker, J. van Donk, 1970, Rev Geophys Space Phys 8: 169)

213

be modeled on the time scale which is appropriate here. The rapid warming which follows maximum cooling, and gives rise to the saw-tooth effect, is perhaps the most puzzling feature. It calls for strong positive feedback within the atmosphere-ocean system. Incomplete mixing of the ocean during deglaciation, and changes in the CO_2 content of the atmosphere, have been proposed as candidates for further investigation.

The very rapid introduction of meltwater, which is evident from the record in the Gulf of Mexico (Fig. 9.7) suggests that some sort of stable layering might have developed in the ocean, during deglaciation. Such an ocean would have had very different dynamics from the present one, developing the conditions favorable for the "climate run-away" effect seen in the record. The problem is of great interest, because the present introduction of large amounts of CO_2 into the atmosphere (Sect. 8.3.6) might generate such a run-away situation artificially.

The Pleistocene cycles, then, appear to become more enigmatic the more we learn about them. Apparently we have found the basic drive — the Milankovitch mechanism — but we are still in the dark about the way the climate machine really works.

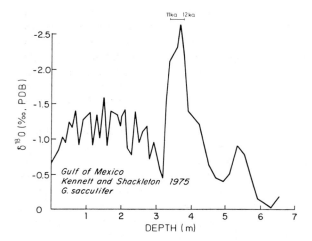

Fig. 9.7. The Gulf of Mexico meltwater event. During deglaciation, fresh water was introduced rapidly enough to change the oxygen isotope composition of surface waters to a considerable degree. The salinity must have been lowered correspondingly. Oxygen isotope stratigraphy from J. P. Kennett, N. J. Shackleton, 1975, Science 188: 147; ^{14}C dating from C. Emiliani et al. (1975, Science 189: 1083)

9.5 Ancient Oceans

In August of 1968, the late Maurice Ewing took the Glomar Challenger on her maiden voyage into the Gulf of Mexico and off the East Coast on the first of a long series of drilling expeditions. Certain knolls in the Gulf were shown to be salt domes, raising the question of the age and the origin of salt in the deep ocean. Two years later, the Glomar Challenger drilled salt in the deep floor of the Mediterranean. Were these seas isolated at some time from the open ocean? Did they dry out, and develop enormous salt pans? Apparently so.

Also on the first leg, the Glomar Challenger drilled chert in the western North Atlantic, at the level of a prominent *seismic reflector* (= acoustic echo from below the sea floor, see Fig. 1.9). The chert was found to be of Eocene age. Later legs found more Eocene chert in other parts of the oceans, notably in the tropical Pacific.

What was different about the Eocene ocean to encourage such widespread formation of siliceous deposits?

These questions are by no means resolved, and they are only a few of the problems raised by the results of deep sea drilling. To give a list, even of just the major ones, with the various hypotheses put forward to explain them would require a book by itself. Let us concentrate here on three important aspects of the history of ancient oceans:

1. The oxygen isotope record since the Cretaceous.
2. The fluctuations of the carbonate compensation depth.
3. The changes in oxygenation of the deep sea.

9.6 Tertiary Oxygen Isotope Record

9.6.1 General Trends. The initial work on the oxygen isotope record of the Tertiary deep ocean was done by C. Emiliani, who identified an overall cooling trend since the Cretaceous, from the increase of oxygen-18 in benthic foraminifera. He deduced that the cooling of bottom waters reflected a similar cooling in high latitudes, since that is where the bottom waters sink today. The overall cooling trend in high latitudes as well as fluctuations superimposed on the trend were shown in an isotope curve by Devereux, who studied shallow water bivalves in New Zealand.

Much more detailed information has now become available through the Deep Sea Drilling Project (see Fig. 9.8). The following is one possible interpretation of these results, not necessarily correct in all aspects.

The oxygen isotopic composition of planktonic and of benthic foraminifera shows separate trends for low latitudes, but similar trends for high latitudes. Thus, the Tertiary cooling trend is indeed largely a high latitude (and deep water) phenomenon. In general, then, temperature gradients must have increased throughout the Tertiary, especially since the end of the Oligocene some 20 to 25 million years ago. We have earlier pointed out that wind speed depends strongly on temperature gradients. If so, winds and their offspring, the surface currents, greatly increased since the Oligocene, as did coastal and mid-ocean upwelling. Direct evidence that this is true is found in the increasing diatom supply, both in the northern North Pacific and around the Antarctic, during the Late Tertiary. Other evidence for fertility increases also exist. For example, radiolarian skeletons become more delicate after the Early Tertiary. Apparently the content of dissolved silica decreased through time, a sign that silica has been vigorously removed by diatom production in upwelling areas.

9.6.2 Eocene-Oligocene Step. The polar regions cooled at the end of the early Tertiary, presumably due to thermal isolation of the Arctic and Antarctic and due

215

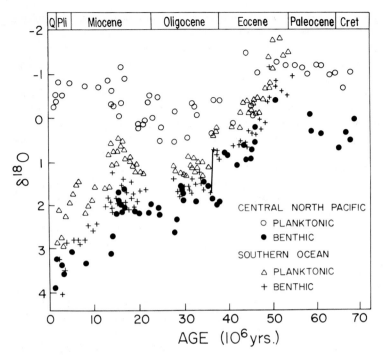

Fig. 9.8. Tertiary oxygen isotope record of the deep sea. Central North Pacific: Data from R. G. Douglas, S. M. Savin (1975, Init Rep Deep Sea Drilling Project 32: 509). Southern Ocean: Data from N. J. Shackleton, J. P. Kennett (1975, Init Rep Deep Sea Drilling Project 29: 743). Note the maintenance of warm temperatures in the surface waters of low latitudes (planktonic, central North Pacific) and the strong cooling at high latitudes and in deep waters since the Eocene (all other points). Cooling seems especially pronounced at the Eocene-Oligocene boundary and after a mid-Miocene warm period near 15 million years ago. (Diagram from H. R. Thierstein, W. H. Berger, 1978, Nature [London] 276: 461)

to positive feedback from snow cover and changing vegetation (increase of *albedo*, Sect. 9.2). The polar front migrated Equator-ward, pushing a late Eocene rainbelt (and thus the temperate regions) away from Antarctic shores. We see the effect on the oxygen isotope values in high latitudes: they quickly move to higher values during the late Eocene, with an especially dramatic change at the Eocene-Oligocene boundary. This is the point at which the water on high latitude shelves is cold enough, and saline enough, to sink and fill the deep ocean basins.

What is the effect of this Eocene-Oligocene boundary event? Obviously, there is an effect on the deep water fauna which has to adjust to the new environment, the cold water sphere. Indeed the evidence shows great changes in the composition of ostracods and other benthos. Less obviously, the chemistry of the ocean must change. For example, CO_2 is suddenly much more soluble in the sea. Hence, it is extracted from the atmosphere. This constitutes positive feedback, probably leading to cooling of the entire planet. Other feedback mechanisms may lead to further cooling. Note the distinct drop in tropical

216

temperatures which is evident from the isotope values of planktonic foraminifera, in Fig. 9.8.

What started the terminal Eocene event? Several hypotheses have been expressed. Perhaps the simplest one is that the late Eocene regression bared the shelf areas sufficiently to produce cooling by albedo feedback (land is brighter than ocean). This set off the Equator-ward migration of the polar front. Eventually a critical position was reached, at which time sink-water formed. Once cold water started sinking, extraction of atmospheric CO_2 and positive feedback from a change in albedo accelerated the process: a run-away effect which lasted for some 100,000 years.

Why did the Oligocene stay cool? Presumably the regression which started in the late Eocene continued into the Oligocene, and sea level was relatively low throughout the period. Bared shelves, of course, would have increased albedo. Also conditions of drought apparently were widespread (cold air does not hold and transport much water). This is reflected in low rates of deep sea sedimentation during the time. Drought makes the land bright. Thus, much of the sun's radiation was reflected into space. The planet stayed cool.

9.6.3 Late Tertiary Changes. Following now the isotopic trends into the Miocene, we see a *rise* of isotopic temperatures in the tropics, and soon after a *fall* in high latitudes. What produced the rise? Why the fall?

The simplest explanation for the rise is transgression. At the beginning of the Miocene the seas again invaded the shelves of the continents. Albedo decreased, the Sun's radiation was accepted to a greater degree and turned into heating the atmosphere and the upper ocean waters. The *contrast* between tropics and polar areas was greatly increased through this process, because snow and ice kept the high latitudes frigid.

The fall is part of the overall cooling trend, accelerated at this time by as yet unknown mechanisms, presumably including a build-up of polar ice.

The remainder of the isotope data shows progressive cooling and growth of ice in high southern latitudes. Finally, in the Late Pliocene (about three million years ago) the northern glaciation set in, and the planetary climate moved into the glacial cycles which we discussed earlier (Fig. 9.9). Again, the Plio-Pleistocene cooling was associated with regression. In addition, mountain building and the drift of land masses northward toward polar areas helped bring about a greater likelihood for snow staying on the ground — the necessary condition for glaciation.

9.7 Fluctuations of the Carbonate Line

9.7.1 Paleodepth Reconstruction. Much like snow which covers the mountains and dissolves in the valleys, calcium carbonate covers the more elevated areas of the sea floor and dissolves at great depths (Fig. 9.10, see also Sects. 8.3.1 and 8.3.2). The lower limit of snow cover, the snowline, is analogous to the *carbonate*

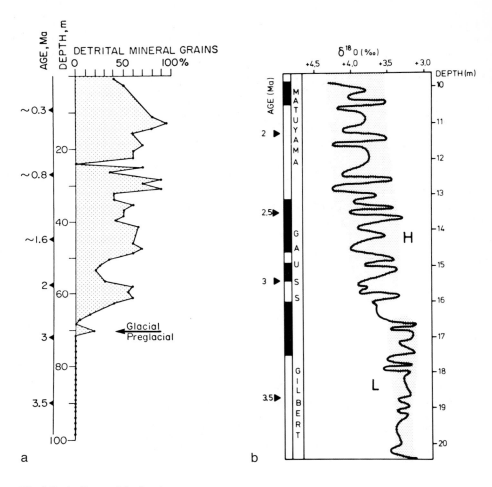

Fig. 9.9a, b. Onset of the Ice Age.
a Sudden increase in the abundance of detrital mineral grains at DSDP Site 116, NW Atlantic, indicating glacial activity on the adjacent continent (W. A. Berggren, 1972, Init Rep Deep Sea Drilling Project 12: 953). **b** Change from a low variability climate *(L)* to a high variability ocean climate *(H)*, as indicated in the oxygen isotope stratigraphy of a piston core from the western equatorial Pacific. The isotopes refer to the benthic foraminifera *Globocassidulina subglobosa* whose present composition is near +3.5‰ at this site. Magnetic stratigraphy controls the time scale. (N. J. Shackleton, N. D. Opdyke, 1977, Nature [London] 270: 216)

line or *CCD*. The position of the CCD varies through time. How exactly does the CCD fluctuate? What can this tell us about the changing chemistry and fertility of the ocean?

Before we can reconstruct the CCD fluctuations, we have to account for the subsidence of the sea floor, down the flanks of the Mid-Ocean Ridge (Sect. 1.5). We use the generalized subsidence curve to find the depth of deposition (Fig. 1.10).

218

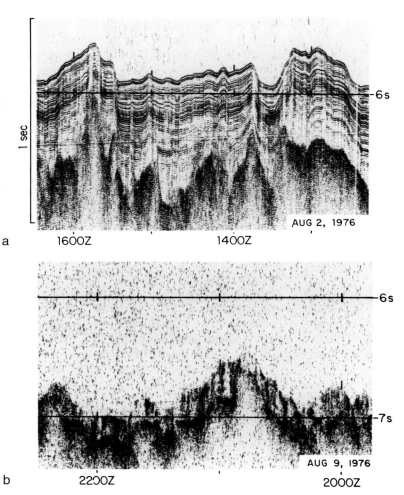

Fig. 9.10a, b. Effect of depth on carbonate sedimentation.
a Seismic profile of well-layered, well-preserved calcareous sediments, western flank of the East Pacific Rise, near the Equator. *6s* is the echo-return time and corresponds to 4500 m depth.
b Similar profile, further west, at greater depth (*7s*≙5250 m). Topography and sediment cover are irregular, and are interpreted as deep sea karst, produced by dissolution of carbonate which migrated downward with the spreading and subsiding sea floor. (From W. H. Berger et al., 1979, Mar Geol 32 : 205)

First, we identify the point which corresponds to the basement age of a given deep sea record, on the subsidence curve of Fig. 1.10. Then we walk back up the curve, the distance corresponding to the age of the sediment for which we wish to determine the depth of depositon. We get a paleodepth estimate, as a difference to present water depth, for the time of deposition. To refine this estimate, we subtract one half of the thickness of the sediment deposited since, to account for both the buildup of the sea floor and for isostatic subsidence.

219

Now we have a method ("backtracking") to reconstruct the CCD fluctuations. In an age–depth diagram, each drill site will appear as a subsidence curve, on which the sedimentary facies (and rates) can be plotted. Fig. 9.11 shows two examples, based on results of Leg 3 and Leg 14 of the Deep Sea Drilling Project (Atlantic). The sediment stack of Site 15 shows carbonate on top of basement, at 140 m in the bore hole, then clay (between about 110 to 100 m), then carbonate again. The CCD was crossed twice. We can see where, when the site is plotted into the age–depth–facies diagram (near 3.3 and 3.5 km depth). With enough such tracks, the diagram can be completed and contoured.

9.7.2 Atlantic and Pacific CCD. Reconstructions of CCD fluctuations differ somewhat depending on what data on ages are taken and what assumptions about subsidence are made. However, it is clear that the CCD stood high in the late Eocene, dropped in the earliest Oligocene, rose in the Miocene when it reached a peak between 10 and 15 years ago, and then fell to its present depth near 4.3 km (Fig. 9.12).

There is an overall similarity in the CCD fluctuations of Pacific and Atlantic: a sign that the chemical climate of the ocean is changing on a global scale. Eocene and Miocene, on the whole, have a shallower CCD than Oligocene and Plio-Pleistocene. A certain parallelism of this pattern with oxygen isotope variations (Fig. 9.8) and with sea-level variations (Fig. 5.18) is obvious. Periods of high sea level are characterized by shallow CCD and warm high latitudes; periods of low sea level have a deep CCD and cold high latitudes (and deep waters).

9.7.3 Causes of CCD Fluctuations. We have already discussed the relationship between paleotemperature and sea level: transgression equals low albedo, hence warming; regression equals high albedo, hence cooling. But why should a relatively warm ocean have a *shallower* CCD than a cold one? Is not cold water *less* favorable to the preservation of carbonate than warm water? This paradox illustrates the fallacy of taking any one factor in the carbonate budget out of context: only when *all other factors stay constant* can we predict what the change of one factor such as temperature would do to the system. Factors rarely stay constant. The closest one can come to such a situation is when the change in one factor is extremely rapid, so that the others do not have time to adjust to the new conditions.

A simple hypothesis linking sea level to CCD fluctuations is the concept of *basin–shelf fractionation*. The shelf, being shallow, is the favored place for carbonate to accumulate, because of the correlation between solubility and pressure. Flooded shelves, then, are carbonate traps, and remove $CaCO_3$ from the ocean so that the deep sea floor starves. Conversely, bared shelves supply carbonate to the deep sea.

However, there are some problems. If the basin–shelf hypothesis were truly sufficient for explaining CCD fluctuations, a deep CCD should mean *more carbonate* in the deep ocean. After all, this is the cornerstone of the basin–shelf argument. Unfortunately for the hypothesis, sedimentation rates were actually at

220

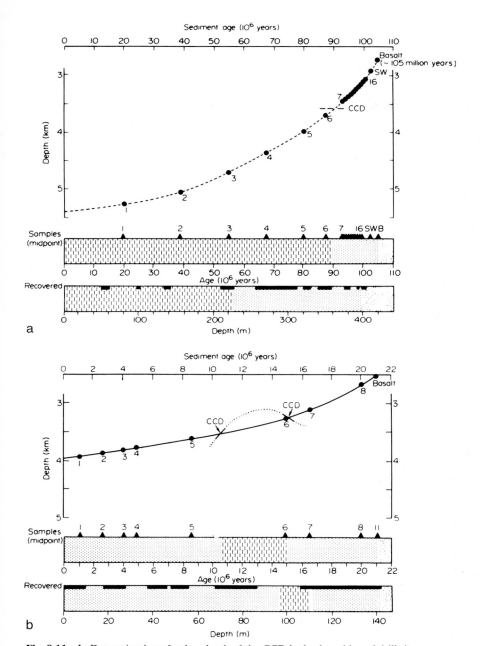

Fig. 9.11a, b. Determination of paleo-depth of the CCD by backtracking of drill sites.
a Subsidence track and sediment sequence of Site 137 Central Atlantic. Lithologic symbols (from left to right at base): *broken lines* clay; *dots* calcareous ooze; *random dashes* basalt. CCD is reached near 3.6 km depth, 90 million years ago.
b Subsidence track and sediment sequence of Site 15 South Atlantic. Symbols as in **a**. The CCD was crossed twice by this track, which suggests an excursion of the CCD toward shallow depths about 12 million years ago. (From W. H. Berger, E. L. Winterer, 1974, Spec Publ Int Assoc Sedimentol 1: 11)

221

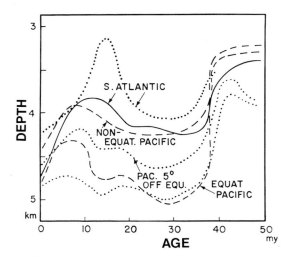

Fig. 9.12. Reconstructions of CCD fluctuations for various oceanic regions. *Solid line* and *dashed lines* reconstructions of Tj. H. van Andel et al., 1977, J Geol 85: 651. *Dotted lines* reconstructions of W. H. Berger, P. H. Roth, 1975, Rev Geophys Space Phys 13: 561. The reconstructions agree in the general pattern of the fluctuations, which appears correlated with sea level changes, on the whole

an all-time low in the Oligocene when the CCD was depressed. *Mass balance alone*, that is, the shifting of carbonate back and forth between shelves and basins, cannot control the CCD fluctuations.

The mass balance hypothesis does not suffice. We must find a more subtle argument, based on *internal cycling* of carbonate within the deep ocean basins. Recall that the biological productivity of the ocean is responsible for precipitating carbonate. Removal of solids from a solution lowers saturation. High productivity, then, results in an undersaturated ocean with a shallow CCD. In this model, we can read the CCD fluctuations as productivity fluctuations, with fertility high in the Eocene and Miocene, low in the Oligocene. There is other independent evidence that productivity was low in the Oligocene, supporting this concept.

CCD fluctuations also are closely associated with the history of *erosion* on the deep sea floor. The dissolution of carbonate itself is a form of erosion, of course. Compilations of sedimentation rates and hiatuses show that both these stratigraphic parameters fluctuated considerably through time (Fig. 9.13). However, the relationships between these fluctuations and those of the CCD are still obscure. One problem is that the hiatus curves and the sedimentation rate curves have artificial elements. For example, a trend toward higher rates (and fewer hiatuses) throughout the Tertiary is simply expected from the fact that older sediments are reached preferentially in areas of low sedimentation rates, by the drill. Also, hiatuses may be "produced" during drilling, when recovery is difficult. This happens, for example, in the chert-rich sediments of Eocene age.

9.8 Fluctuations in Oxygenation of the Deep Ocean

9.8.1 A "Stagnant" Ocean? When the Deep Sea Drilling Project first recovered mid-Cretaceous sediments in the Atlantic Ocean, it was found that some of these were very rich in organic matter, indicating either high organic supply to the sea

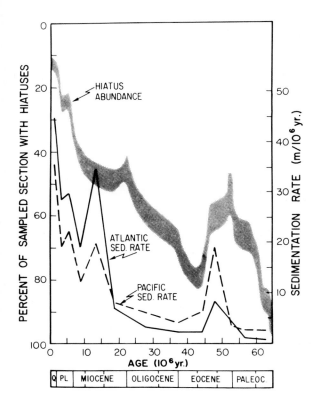

Fig. 9.13. Tertiary history of deep sea sedimentation rates (from T. A. Davies et al., 1977, Science 197: 53) and of hiatus abundance (from T. C. Moore, G. R. Heath, 1977, Earth Planet Sci Lett 37: 71). Note the general trend toward thicker and more complete sections (=higher rates) since the early Tertiary. In part this trend is due to increased erosion on the continents, in part it is an artifact (see text). (Diagram from W. H. Berger, 1979, in M. Talwani et al. [eds], 1979, Maurice Ewing Ser, vol III. Am Geophys Union, Washington DC)

floor, or reduced losses of organics due to low oxygen content in deep water at the time, or both (Fig. 9.14). Regionally the oxygen content apparently dropped low enough to prevent burrowing organisms from establishing a mixed layer on the sea floor. Thus, laminations are common in deep sea sediments of this time.

The various indicators of low oxygen conditions are concentrated during certain well-defined periods which may have been quite brief ("anoxic events"). They occur typically in the Atlantic, although some have been found in the Pacific also. What is the significance of these anaerobic and near-anaerobic deposits? What are the implications for the chemistry and fertility of the ocean at the time? For climate?

9.8.2 A Search for Causes. To try to answer these questions, let us first search for modern analogs of deposition under low oxygen conditions (see Sect. 7.6.3). There are two situations: the partially restricted basin with estuarine circulation, and the open ocean continental slope, where it is intersected by a strong oxygen minimum (Fig. 9.15). The Baltic and the Black Sea are examples for the first, the Gulf of California and the Indian continental slope of the Arabian Sea for the second. Common to both is a high supply of organic matter relative to the oxygen supply from the deep water into which the organic matter falls. To expand the

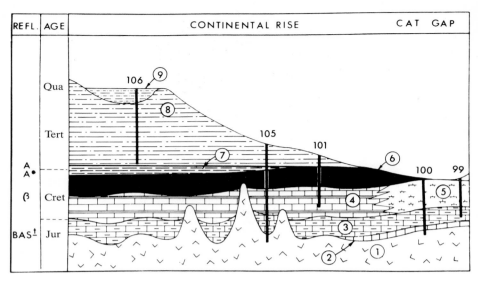

Fig. 9.14. Cretaceous organic-rich sediments in the continental margin off the U.S. East Coast. The diagram is highly schematic, and is based both on drilling and seismic profiler records. *1* Basalt; *2* greenish-grey limestone; *3* red clayey limestone; *4* white and grey limestone; *5* calcareous ooze and chalk; *6* black clay; *7* multicolored clay; *8* hemipelagic mud; *9* terrigenous sand and clay. Age: *JUR* Jurassic; *CRET* Cretaceous; *TERT* Tertiary; *QUA* Quaternary. (From Y. Lancelot et al., 1972, Init Rep Deep Sea Drilling Project 11: 901)

likelihood of obtaining "black" deposits, then, we can increase productivity while keeping the oxygen supply constant, or decrease the oxygen supply, keeping productivity constant.

Which of these conditions applies to the mid-Cretaceous? There is no evidence for high productivity in the mid-Cretaceous, either from sedimentation rates or from the type of sediment delivered. Thus, it has to be a low supply of oxygen which is the crucial factor.

We can readily appreciate why the oxygen content of deep waters was low: the temperature was relatively high. At present, the ocean has a temperature of between 0° and 5° C, except for a thin upper layer. Saturation values for oxygen are near 7.5 ml/l. Typical actual values in the deep ocean are between 3 and 5 ml/l, that is, about 3.5 ml/l lower, because of oxygen consumption through decay. For the Cretaceous ocean, isotopic measurements suggest deep water temperatures close to 15° C. For water temperatures between 15° and 20° C the oxygen saturation is near 5.5 ml/l — 2 ml less than for the present cold deep water. We might expect, then, a typical value near 2 ml/l for the deep Cretaceous ocean, after subtracting a loss to account for decay. Under these conditions, any above-average loss of oxygen in an ocean basin with estuarine-type circulation (Fig. 7.12) would make the basin susceptible for developing regional oxygen deficiency. Unusually high supply of *terrigenous* organic matter to such a basin could further reduce the oxygen supply.

224

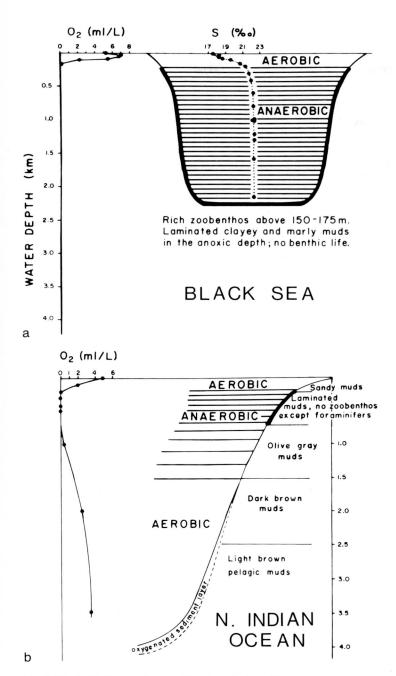

Fig. 9.15a, b. Modern analogues of black shale deposition.
a Black Sea, with strong salinity stratification, preventing vertical overturn and hence supply of oxygen to the deep layers.
b Intersection of an oxygen minimum layer with a continental margin, in the northern Indian Ocean. The minimum layer is a worldwide phenomenon; it is reinforced in areas of high production. (After J. Thiede, Tj. H. van Andel, 1977, Earth Planet Sci Lett 33: 301)

9.8.3 Regional Effects, and Hydrocarbon Deposits. Can the entire ocean become anaerobic? Probably not for long if at all. There is a negative feedback in the system which prevents the ocean from becoming anaerobic altogether: the organic-rich sediments tie up nutrients, and thus deposition under low-oxygen conditions helps reduce fertility, which decreases the supply of organic carbon, and hence the demand on the available oxygen.

Why does the Atlantic have more "black" deposits than the Pacific, in the Cretaceous?

This pattern suggests that the Atlantic had an estuarine circulation with respect to the Pacific, just the reverse of the present situation. The possibility also must be considered that the deep Atlantic collected heavy saline waters from surrounding shelves. In a deep "salt lake" covering the ocean floor, oxygen would soon be used up by combustion of some of the organic matter falling into it. Conditions for the preservation of organic matter would have been ideal under these circumstances.

The exact conditions under which the "black" deposits developed are very much a subject of debate. The outcome is of economic interest. Many important hydrocarbon deposits are thought to be derived from organic-rich sediments of Cretaceous age. Can we predict, from paleoceanographic principles, where conditions should have been just right for the formation of petroleum? Perhaps not quite yet. The origin of source rocks is just one problem of many: diagenetic alteration, migration of hydrocarbons, and trapping is necessary before we arrive at a petroleum reservoir. We shall return to the question of petroleum formation in the following chapter, on resources of the ocean floor.

10 Resources from the Ocean Floor

10.1 Types of Resources

The ocean floor contains energy sources (petroleum and gas) and raw materials (sand and gravel, phosphorite, heavy metal ores). Also, the sea floor is used as a dump site for waste, which represents a considerable economic value. In terms of dollars and cents, energy (hydrocarbons) is the most important resource, while (at present) raw materials are of regional importance only. Nothing as yet has been gained from deep sea ores, although they are of great scientific interest and are *potentially* valuable. We shall briefly treat the geologic background to such resources here, with some mention of the economic and the political problems associated with the use of the sea floor.

10.2 Petroleum Beneath the Sea Floor

10.2.1 Economic Background. The present high standard of living of the industrial countries, unprecedented in all of history, is largely dependent on the availability of large amounts of cheap energy. In the U.S.A. and elsewhere, petroleum delivers about one half of this vital resource (in 1977, North America 48%, Western Europe 56%). In the U.S.A., about half of the oil is imported (this proportion is decreasing), in Germany 96%, in France 99%, and in Japan even 100%. These imports are becoming more and more expensive, reflecting the strong and growing demand, and the realization by the sellers that their goods cannot last forever. Since demand is growing while oil production is leveling off, oil prices are rising rapidly, and petroleum from the sea floor is becoming increasingly profitable, despite the considerable cost of exploration and production. In 1978, world production was about three billion tons of petroleum and 1.4 trillion cubic meters of natural gas. Around 20% came from offshore sources. (The U.S.A. consumed almost 30% of the world's oil supply in 1978). Total potential resources and even reserves (=producible petroleum under prevailing economic and technological conditions) are hard to estimate. World *resources* may be 200 to 300 billion tons of oil and 200 to 250 trillion cubic meters of gas, and world *reserves* 100 billion tons and 70 trillion cubic meters. Expressed in static life expectancy, reserves for oil are good for 33 years (100 billion tons/3 billion tons per year) and for gas, 50 years (70 trillion cubic meters/1.4 trillion cubic meters per year). U.S. oil reserves were estimated between 26 and 76 billion tons in 1972 and recently around 13 billion tons only. Hence, static life expectancy would be fifteen years only, if all U.S. consumption were to be supplied from U.S. reserves.

In the past, reserve estimates have been revised upward with some regularity. In 1950, world reserves were estimated at 10.5 billion tons, while consumption was 0.52 billion tons, that is, the static life expectancy was 20 years. In 1960, these figures were 40/1.05/38, and in 1977, about 100/3/33. So far, then, new discoveries have kept up with increasing use. At some point, of course, new discoveries will fall behind. Some experts think that this is happening right now.

10.2.2 Origin of Petroleum. Petroleum is a complex mixture of hydrocarbons and other organic compounds originating from organic matter produced both on land and in the sea, but largely from marine plankton. Therefore, essentially, petroleum is stored fossil solar energy. But this storage system is extremely inefficient. At present only 0.23% of solar radiation is used for photosynthesis (the basis of organic production). During Earth's history, less than 0.1% of the organic production was preserved in sediments as organic carbon compounds. Only about 0.01% of this organic matter in sediments became concentrated in oil and gas fields. Why so little?

The reasons are manifold: (1) The organic matter captured within the sediment must be converted to fluid petroleum by thermochemical processes, requiring a blanket of sediments more than 1,000 m thick and temperatures of 50° to 150° C. If temperatures become too high, oil is cracked to natural gas (Fig. 10.1). (2) Petroleum must migrate from organic-rich source-rock sediments to porous and permeable reservoir rocks such as sandstones or vuggy limestones in response to compaction pressure and gravity (oil is lighter than interstitial waters) (Fig. 10.2). (3) Reservoirs must be big enough to be of interest, and they must trap petroleum with impermeable cover rocks such as thick shales or evaporites. Otherwise the more volatile hydrocarbons escape to the surface. Examples of lost petroleum are the pitch lakes in Trinidad and Iraq, where oil has leaked through the surface, and the more volatile parts have evaporated into the atmosphere. The La Brea tar pits in Los Angeles are a familiar illustration of the process. About 200 natural marine oil seeps are reported worldwide. Petroleum seepage into the marine environment has been estimated as 0.6 million tons per year (a little less than the amount of oil estimated to be spilled at sea). (4) These several petroleum-forming processes must take place within the correct time-frame: each process needs to complete its turn in the proper sequence. When everything looks just right, the drill-hole may yet be "dry", because the *timing* in the interplay of the natural processes was wrong.

10.2.3 Where Offshore Oil is Found. The classic area of offshore oil fields is in the Gulf of Mexico. The conditions of entrapment are much like those onshore. Marine sediments overlie salt, which is gravitationally unstable, being less dense than the overburden. It pushes up in plumes, making the so-called salt domes The upturned sedimentary strata butting against the salt provide traps for petroleum (Fig. 10.2). Offshore production in the Gulf has been on the order of 250 million barrels (=34 million tons) per year (7% of total U.S. Production).

228

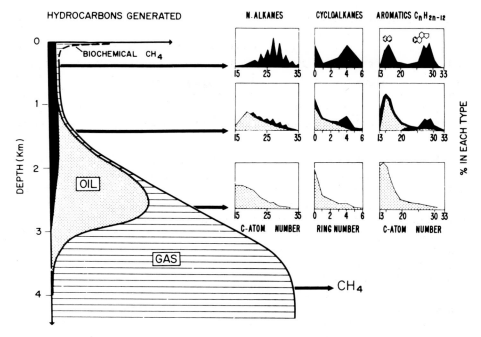

Fig. 10.1. Origin of petroleum. After burial of the organic-rich sediment containing the source material *(black)*, carbon compounds (chains and rings) are produced. In the zone above 1 km depth, the numbers of C-atoms tend to be large within the organic molecules. In the oil-zone, compounds with fewer C-atoms (and more H-atoms) are produced by a cracking process, due to elevated heat and pressure. Under a large overburden, gas is produced (CH_4=methane). (From B. P. Tissot, D. H. Welte, 1978, Petroleum formation and occurrence. Springer Berlin Heidelberg New York)

Another area from which offshore oil has been produced for a long time is the Continental Borderland off Southern California. There, oil originates from Miocene organic-rich marine strata originally formed under conditions of upwelling and oxygen deficiency. The oil is trapped in sandy layers abutting against faults, hence there is considerable natural seepage into the ocean, as at Coal Oil Point near Santa Barbara and at the shores of Santa Monica Bay.

The newest highly promising areas for offshore petroleum are on the Arctic shore of Alaska where large-scale onshore recovery is already proceeding (Prudhoe Bay, estimated reserves 1.5 billion tons). The drawback for offshore exploitation here is that climatic conditions are highly inclement.

In recent years, large reserves of oil and gas were discovered in the thick sedimentary deposits of the North Sea, with Britain taking the lion's share. The North Sea is the most important hydrocarbon province of Western Europe. Recoverable petroleum reserves are estimated at three billion tons (United Kingdom 2.2, Norway 0.8, Denmark 0.06) — just about one year's worth of present world consumption. Natural gas reserves are about 2400 billion cubic meters (almost two year's world consumption). In 1978, production of oil was 71 million tons (53 from United Kingdom waters) and 67 billion cubic meters of natural gas (46 United Kingdom, 14 Norway, 7 Netherlands).

229

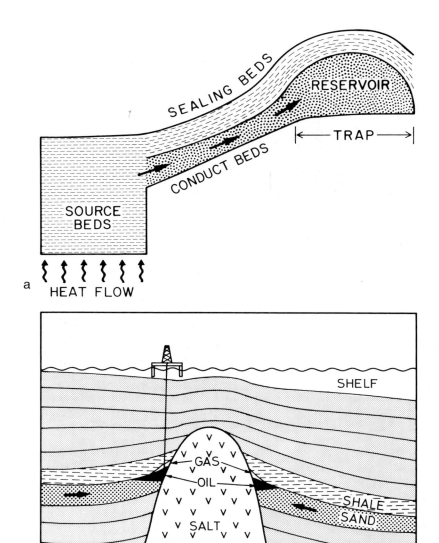

Fig. 10.2a, b. Prerequisites for petroleum accumulation. **a** Basic ingredients of a petroleum reservoir system, showing migration from source beds (shales rich in organic matter) into reservoir (porous rock, e. g. vuggy reef limestone or sandstone) **b** Salt dome tectonics as an example for trapping conditions

Exploration started in the early 1960's, and offshore production began in 1967 for gas and in 1971 for oil. The technical difficulties in recovering hydrocarbons in this "fiercest of all seas" are enormous. Blinding fog, swift and drastic changes in weather, wind speeds of more than 160 km per hour, and waves frequently 30 m high, are dangerous even for the huge platforms. (In March 1980

230

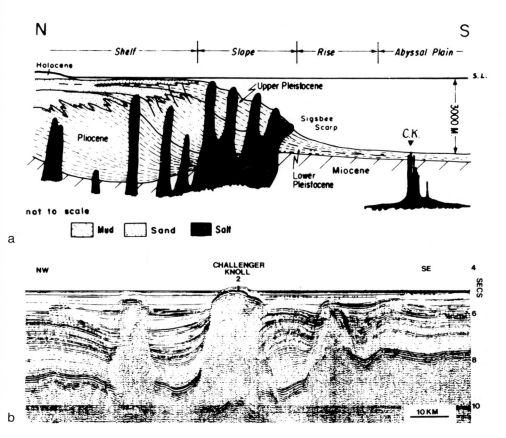

Fig. 10.3a, b. Salt plumes in the Gulf of Mexico. **a** Schematic profile of Pliocene-Pleistocene sediment wedge intruded by early Mesozoic salt deposits. Intrusion of salt in the Sigsbee Abyssal Plain produces the Challenger Knoll. (From C. J. Stuart, C. A. Caughey, 1977, Am Assoc Pet Geol Mem 26: 249; modified)
b Seismic profiler record from Sigsbee Knolls region. Challenger Knoll was drilled in Hole 2 of Leg 1 of the Deep Sea Drilling Project, establishing its nature as a salt dome feature. Time scale is two-way reflection time (sea floor=4.5 s ≙ 3400 m). (From J. L. Worzel, C. A. Burk, 1979, Am Assoc Pet Geol Mem 29: 403)

in the Ekofisk field, the Norwegian platform Alexander L. Kielland collapsed in stormy weather, a catastrophe leaving over a hundred people dead).

In the south, a broad belt of gas fields stretches east–west from Germany to southern England (Fig. 10.4 and 10.5). Gas migrated from Carboniferous coal measures to porous Lower Permian sandstones and is sealed by overlying Zechstein (upper Permian) evaporites. Oil fields, however, are concentrated around a north-south rift in the northern North Sea which originated in the first stages of opening of the northern North Atlantic. The major faults bordering the rift lie on either side of the political median line between United Kingdom and Norway. In the south the oil occurs in Cretaceous chalk reservoirs, next to salt

231

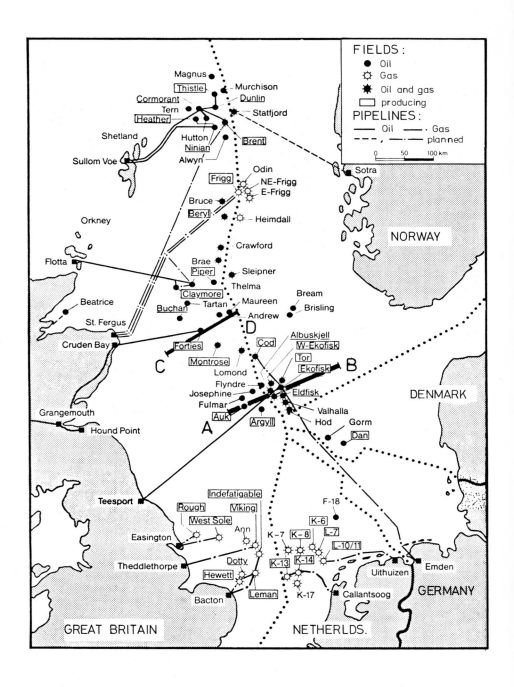

Fig. 10.4. Map of the North Sea showing the location of major oil and gas fields, pipelines and national shelf boundaries. *A-B* and *C-D* refer to cross sections in Fig. 10.5. (From K. Harms, 1978, Oel, Mineralölwirtsch H 10: 267)

232

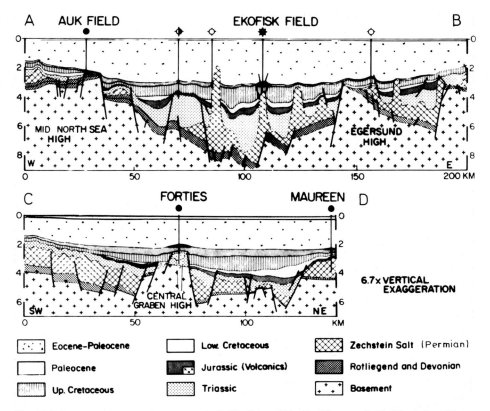

Fig. 10.5. Structural cross-sections across Auk-Ekofisk and Forties-Maureen fields, North Sea. (From P. A. Ziegler, 1977, Geo Journal 1: 7)

domes. The Brent and Piper fields are developed on Jurassic sandstones on the crests of horsts and tilted blocks. The Forties and other fields get oil from basal Tertiary sands.

10.2.4 Present Oil Exploration. As already mentioned, one of the prerequisites of petroleum formation and migration is an elevated temperature. A cover of roughly 1 to 2 km of sediment is required — with some possible exceptions in areas with high heat flow or in sediments where large time spans for conversion are available. Most of the ocean bottom, roughly 80% or even 90%, offers no chances for exploration because it is too young and the sediment carpet is too thin (Fig. 10.6). The most promising prospects for petroleum concentrations are continental shelves and slopes, and small ocean basins with thick sediments produced by high accumulation rates and which are rich in organic matter (e.g., the Gulf of Mexico, the Caribbean, the Mediterranean including the Black Sea, the Bering Sea, Okhotsk, Japan, South China Seas and the Indonesian Archipelago). The enormous sediment wedges of the continental rises bordering

233

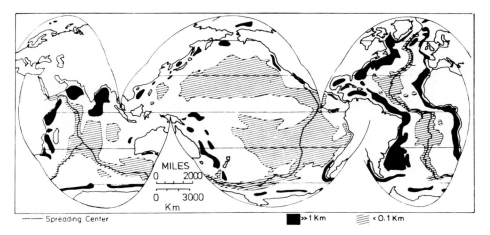

Fig. 10.6. Regions where Cenozoic sediments are well over 1 km thick. Only areas where source beds are deeply buried will produce petroleum (see Fig. 10.1). (From W. H. Berger, 1974, in C. A. Burk, C. L. Drake (eds), 1974. The geology of continental margins, Springer, Berlin Heidelberg New York)

the large ocean basins, although rich in organic matter in places, may or may not contain suitable reservoir rocks; this question is being explored.

The shelf off the U.S. East Coast has recently become the site of intensive exploration, and leases have been granted in places, after much opposition from citizen's groups concerned about the possibility of oil pollution. As shown by occurrences off California (Santa Barbara oil spill, 1969), in the North Sea (Ekofisk oil spill 1977), or in the Gulf of Mexico (Ixtoc-1 oil spill 1979), such concern is not without basis. The problem is a political one of profit and risk distribution; effects of spills on the ocean as a whole cannot be shown to be damaging (but neither is it known whether spills are innocuous). Regional damage, of course, can be extensive.

A few years ago, off the U.S. East Coast, the Baltimore Canyon Trough area (Fig. 10.7) was regarded as the best prospect of the outer continental shelf, because of a sediment thickness of more than 10 km and favorable structural and other conditions. Since 1976, 15 wells were drilled, but 13 were dry, while two showed gas, but no oil. By 1973, 40 wells had been drilled in the Scotian Basin off Eastern Canada, but no economically producible fields have been found.

In addition to the shelves, the upper continental slopes are interesting areas for petroleum exploration. Their sediments were formed in the realm of fertile coastal waters (including deltaic conditions). Hence carbon content of the sediments is high (Fig. 10.8). Also, an oxygen minimum develops in many areas on the upper slope (see Fig. 9.15), and favors burial of organic matter. We have briefly mentioned the Mesozoic structures identified as salt domes in the eastern South Atlantic, off Africa (Sect. 2.3, Fig. 2.4) where organic-rich Cretaceous sediments were found by deep drilling. Clearly, from the Gulf of Mexico experience, petroleum may be expected in such a situation.

234

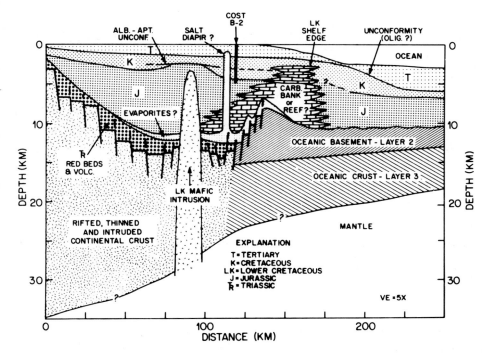

Fig. 10.7. Schematic cross section Atlantic City — Baltimore Canyon Trough off New Jersey. Based on seismic profiles by the U.S. Geological Survey. (From J. A. Grow et al., 1979, Am Assoc Pet Geol Mem 29: 65)

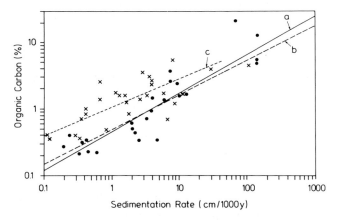

Fig. 10.8. Organic carbon content in marine sediments. The content rises with sedimentation rate: it doubles when sedimentation rate quadruples *(Line a)*. *Line b* regression on dots (Müller and Suess). *Line c* regression on crosses (Heath et al.). (Data and regressions from G. R. Heath et al., 1977, in N. R. Andersen, A. Malahoff (eds), The fate of fossil fuel in the oceans. Plenum Press, New York, and from P. J. Müller, E. Suess, 1979, Deep Sea Res 26A: 1347)

Regardless of the promise of certain continental slope regions, activity on the shelves, even in regions with heavy seas or in polar areas, is going to dominate petroleum recovery from the sea floor in the foreseeable future. Expenses of drilling increase rapidly beyond the shelf break because platforms either have to have extremely long legs, or they have to be positioned dynamically, that is, by constant calculated application of power to several thrusters. Also, safety problems rapidly increase with depth. Beyond the continental slopes, in the deep sea proper, where sedimentation rates are low, the chances of finding oil rapidly diminish.

10.3 Raw Materials from Shelves

10.3.1 Phosphorite. The naturalist on HMS Challenger, John Murray (unlike his modern counterparts) became a wealthy man as a result of joining an oceanographic expedition, due to his interest in a mining venture which resulted from the discovery of phosphorite on Christmas Island in the western equatorial Pacific. What is more, the British Crown's Treasury recovered the expenses of the expedition, through the taxes derived from the phosphorite produced.

The origin and distribution of modern phosphorite was briefly discussed (Sect. 3.8.2): it is commonly found in shallow sea floor, below former or present regions of upwelling. It is both of biogenic origin (phosphatic hard parts such as fish debris and crustacean carapaces), and of hydrogenous origin (replacement of calcium carbonate, precipitation from interstitial waters). Normally, offshore phosphorites contain less than 25% P_2O_5. Weathering processes on land concentrate P_2O_5 up to 30% to 40%. Such residual phosphorites are mined, for example, in Florida and Morocco, largely for fertilizer, but also as a source of phosphorus in the chemical industry. Evaluation of resources as reported from off California, western South America, South Africa, or on the Chatham Rise east of New Zealand, will depend on the price policy of producers on land and on the domestic demand and supply situations.

10.3.2 Shell Deposits. Calcareous shell deposits were or are dredged in places as raw material for calcium carbonate, and also for building roads. For example, oyster shells have been mined in San Francisco Bay, for use in making cement, and in Galveston Bay in the Gulf of Mexico for calcination and reaction with seawater, to extract magnesium. Dredging of shells hinders the growth of benthic organisms and affects adversely the productivity of the sea bed. Conflicts commonly arise where dredging and fishing activities overlap.

With the advent of worldwide souvenir markets and of face-mask diving, the collecting of shells and corals has become an important source of income for islanders of the Pacific and other coastal peoples. Not surprisingly, attractive and rare species suffer considerable depredation from such collecting.

10.3.3 Placer Deposits. Concentrations of heavy minerals and ore particles on beaches and in estuaries are locally mined for metals such as titanium, gold, platinum, thorium, zirconium, and valuable minerals such as diamond. Seventy percent of the world production of zirconium is being extracted from placer deposits off East Australia. Diamonds are found in beach deposits of Southwest Africa, as well as offshore. Magnetite is being mined from beach placers in certain areas of Japan and New Zealand. In the U.S.A., gold has been mined from beach deposits near Nome, Alaska; thousands of tons of ilmenite ($FeTiO_3$) were extracted at one time from Redondo Beach, California; beaches in western Oregon yield chromite and other heavy minerals as well as gold and platinum; titanium minerals are taken from beach sands along the eastern Florida coast.

How do placers originate? The process of concentrating heavy particles on the beach has much in common with panning for gold — water motion works on the different settling velocities of the particles and on their different sizes, to separate heavy from light, large from small.

Terrigenous beach sands commonly consist of more than 95% quartz, tropical ones of calcareous grains. Volcanic rocks and other igneous sources supply minerals which are considerably heavier than quartz or calcite (density of 2.65 and 2.70). These minerals have densities of greater than 2.85 g/cm^3. The "heavy minerals" are ubiquitous, usually making up a few percent of the sand. The back-and-forth of the waves washing over the beach face can concentrate them

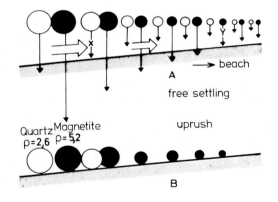

free settling

uprush

Quartz Magnetite
$\rho=2{,}6$ $\rho=5{,}2$

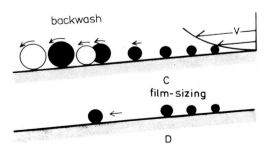

backwash

film-sizing

Fig. 10.9A–D. Origin of heavy mineral placers, schematic. **A** suspended sediment particles are washed on shore, settle to the floor, large and heavy ones (shown solid) first. **B** resulting grain association shows enrichment of large and heavy particles on the beach face. **C** back wash rolls away large particles (greater velocity away from sediment interfac, V, to the *right*). **D** resulting association consists of medium-size well-sorted heavy minerals. (From E. Seibold, 1970, Chem Ing Tech 42: A 2091)

Fig. 10.10. Beach placers. The beach south of Quilon (SW India) has a heavy mineral deposit on the top part (where the boat rests). During SW monsoon, high waves sort the sand as shown in Fig. 10.9, and produce layers of "black sand" *(inset)*. The inset profile is 20 cm high. (Photo E. S.)

greatly (Fig. 10.9 and 10.10). Under the microscope one recognizes minerals such as ilmenite (iron-titaniumoxide), rutile (titaniumoxide), zircon (zirconiumsilicate), monazite (phosphate, containing cerium and thorium).

Thick placers of heavy particles only form in the beach zone. Offshore placers, therefore, were formed when the sea level stood lower. The gold deposits off Nome, Alaska, are a case in point. Glaciers of the last ice age brought gold-containing debris from the hinterland and dropped it on the shelf. This morainal material was then worked up by the surf, as the sea level rose.

River deposits may contain heavy mineral concentrations, too, such as cassiterite (SnO_2) in Thailand, Malaysia, or Indonesia. In these areas former river beds on the shelf, which are now in up to about 100 m water depth, are of potential economic interest and are exploited offshore in shallower water.

10.3.4 Sand and Gravel. Considerable amounts of sand are taken locally for building roads and houses, as well as for coastal protection (dams) and for fill. However, the main use of beach sand is for building sand castles and for plain enjoyment. Recreation is one of the most important uses of beaches — by far exceeding mineral extraction in business value. Beaches commonly are eroded by winter storms (Sect. 4.2.2); in some areas sand is brought in from offshore to replace the eroded material.

238

Gravel only rarely reaches the sea, unless high mountains abut against it, or unless glaciers bring morainal debris. Such debris, when washed by waves or rivers on the bare ice age shelves can then yield gravel. In the Baltic and in the North Sea, for example, gravel is mined for a filler for concrete.

10.4 Heavy Metals on the Deep Sea Floor

10.4.1 Importance of Manganese Deposits. The rich metal deposits on the deep sea floor have been a favorite topic of discussion among marine geologists ever since the *manganese nodules* (better: *ferromanganese concretions*) were discovered by the Challenger Expedition a century ago and were shown to be rich in copper, cobalt, nickel and other heavy metals (Fig. 10.11). The origin of these deposits is still unexplained, meaning that we do not understand the chemical processes on the sea floor, as regards these and other metals. The abundances of these metals in seawater are extremely low, so that their geochemical cycles cannot be followed by measuring concentration changes in seawater. Instead, their distributions have to be mapped on the sea floor, and in the interstitial waters, to obtain clues about their sources and paths of migration.

The amount of nodules on the Pacific Ocean floor seems to be in the neighborhood of 100 to 200 billion tons. What is the economic value of these deposits? Right now, close to zero. It is too expensive and (because of involved international legal problems) too risky to mine them, transport the material back to shore, extract the wanted heavy metals, market them, and still make a profit. Nevertheless, there is a *potential* value, if prices of copper, nickel and cobalt risse high enough. This potential value is a bone of contention among UN members: those nations technically unable to mine the material wish to make sure that those able to do so must share the profits, if any should materialize. In addition, metal-exporting countries are apprehensive about the potential competition. The resulting concept of the deposits as a "common heritage of mankind" has slowed development of the resource, since venture capital is hesitant to take the risk both of failure and of an unknown degree of taxation in case of success. Eventually, however, the resource will probably be exploited; meanwhile, active exploration is being carried out by Japanese, French, German, Soviet, and U.S. expeditions, especially in the central Pacific.

10.4.2 Nature of Manganese Deposits. What are the ferro-manganese deposits like? Where do they occur? How much of the valuable trace metals do they contain? How did they originate? Except for the last one, the answers to these questions are reasonably well known.

The *appearance* of the manganese deposits varies. Nodules come in sizes of 1 to 10 cm and look much like small potatoes except for being black. The surface can be smooth or rough. Not all deposits are nodules: some are crusts several centimeters thick, others are pavements covering the sea floor in areas of active currents. The nodules are rather porous. They are easily crushed, which

239

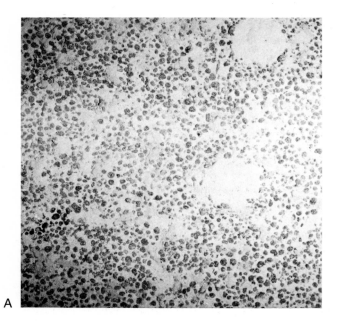

Fig. 10.11A, B. Manganese nodules. **A** View ($\sim$10 m^2) of nodule-covered deep sea floor, central tropical Pacific (Photo Metallgesellschaft Frankfurt)
B Manganese nodules recovered on Challenger stations in the Central Pacific. *a* nodule with upper smooth and lower uneven surface (Station 274, 5000 m). *b* sliced section (parallel to the sea bottom) showing internal layering and nuclei (Station 253, 5700 m). (From J. Murray, A. F. Renard, 1891, Report on deep-sea deposits. H. M. S. Challenger, 1873–1876. Reprinted 1965 by Johnson Reprint, London)

should facilitate on-board processing and chemical extraction of metals when the time comes. On cutting the nodules, one notes a concentric structure. In the center there is commonly a core of altered volcanic material. Fragments of older nodules, bones, or shark teeth also can serve as core. From the age of such cores

240

it is quite obvious that the deep sea nodules must grow very slowly, a few millimeters per million years, at most.

The manganese nodules occur in areas of low sedimentation rate: because of their slow growth they would soon be covered up in regions of high sediment supply (Fig. 10.12).

Calcareous ooze accumulates at about 10 m per million years, 1,000 to 10,000x faster than nodules; hence no nodules develop (excepting "micronodules" and encrustations on shells of foraminifera in many areas). Ferromanganese is deposited here also, but is greatly diluted with carbonate. Brown pelagic clay (Red Clay) accumulates at less than 1 m, up to about 2.5 m, per million years, the higher values applying in the Atlantic Ocean. Surprisingly, nodules have time to grow at the surface under these conditions. Apparently they are moved about sufficiently, through the activity of benthic organisms, to stay on top of the clay. In fact, such movement may be necessary to keep them round rather than mushroom-shaped, and to prevent them from coalescing into a pavement. If we could mount a time-lapse camera on the sea floor, taking a picture every 100 years for 10,000 or 100,000 years, what a spectacle of dancing nodules we might see!

The distribution of nodules is patchy. For example, in the eastern central Pacific, records were taken with bottom-near television cameras, for hundreds of kilometers. For 5% of the records the sea floor was covered to more than 50% by nodules, that is, up to 25 kg of ore per square meter. Conversely, 5% of the records showed sea floor free of nodules. Elsewhere the cover varied, sometimes considerably, even over distances of only 50 m. The reasons for patchiness are not clear; in part it may be due to alternating burial and uncovering of nodules below slowly moving clay carpets, somewhat akin to dune migration.

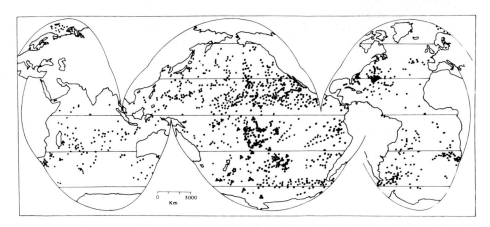

Fig. 10.12. Areas from which ferromanganese concretions have been reported. The richest areas are in the north-equatorial Pacific. (From W. H. Berger, 1974, in C. A. Burk, C. L. Drake [eds], 1974. The geology of continental margins. Springer, Berlin Heidelberg New York; mainly after D. R. Horn [ed] 1972, Ferromanganese deposits on the ocean floor. Harriman, New York)

10.4.3 Origin of Manganese Nodules. How manganese nodules originate is not agreed upon. Questions about the origin of ferromangese concretions commonly address themselves to any number of the following aspects of ferromangese distributions: (1) the ultimate source — for example, weathering of continental or oceanic rocks or sediments or exhalations from volcanic or hydrothermal vents. (2) The immediate source, that is, the surrounding seawater or the mobilization of ferromanganese and other metals (by reduction) within sediments and diffusion to the interface. (3) The mode of extraction from the source; for example, differential weathering or hydrothermal reactions. (4) The mode of transport into the ocean — that is, whether delivery is as solute, mineral, or coating on mineral. (5) The transport and migration processes within the ocean — i.e., the question of how ferromanganese gets to its place of deposition, as colloid or other suspensate, in solution, or locked up in organic matter and plankton shells. (6) The physicochemical and biochemical methods of precipitation and their variation through time. (7) The processes of redeposition, physical and chemical, including movement of nodules, scouring of paved sea floor, as well as dissolution and reprecipitation of ferromanganese.

The first of these problems, that of ultimate sources, was the one originally raised by Murray and Renard (1891) and has dominated much of subsequent research. The discussion since have revolved mainly around two types of arguments: (1) The association of ferromanganese concretions with volcanogenic material versus association with other matter, arguments which are definitely unconvincing. (2) The regional variations of trace element composition and of morphologic characteristics, compared with expectations for volcanic or continental origin.

The problem is, of course, that we do not know what to expect, and to what degree the distributions depend on the ultimate origin at all. There is, however, an unusually high rate of sedimentation of (disseminated) iron and manganese in the region of the East Pacific Rise, centered on the crest. This suggests that input of Fe and Mn from the hydrothermal activity at the crest is indeed important.

In the various discussions about the origin of ferromanganese, the *trace element* content, as well as the *iron-to-manganese* ratio play an important role. Typical values for Pacific, Atlantic, and Indian Ocean are given in Table 10.1. Note that the Mn/Fe ratio is greater in the Pacific than in the Atlantic. Also, the content of trace elements in Pacific nodules, on average, is about twice that of Atlantic ones.

What factors control the Mn/Fe ratio? Which ones control the trace element content?

The Mn/Fe ratio, on the whole, increases with the degree of oxidation and with depth — the more "deep sea" character the nodules have, the more manganese they contain. Shallow water ferromanganese concretions (e.g., on continental slopes) are generally iron-rich. In the economically interesting manganese zone north of the central equatorial Pacific, Mn/Fe ratios go as high as ten. High manganese content appears to be favored by both high biogenous sediment supply and low rates of accumulation. Under these conditions, presumably, the carrier material dissolves but leaves its content of trace elements. Also, in the "manganese zone" in the Pacific, the underlying sea floor consists of

Table 10.1. Metal contents of manganese nodules ($\%$) (mainly after J. S. Tooms, 1972, Endeavour 31: 113)

	Average			Extreme values (Pac. & Ind. Ocn.)	
	Pacific	Indic	Atlantic	max.	min.
Mn	17.2	14.9	13.6	34.00	5.41
Fe	11.8	14.6	15.5	26.32	4.36
Ni	0.63	0.38	0.33	2.00	0.13
Co	0.36	0.31	0.24	2.57	0.045
Cu	0.36	0.17	0.16	2.5	0.028
Pb	0.047	0.053	–	0.51	0.0046
Ba	0.20	0.16	–	1.58	0.018
Mo	0.036	0.031	–	0.080	0.0087
V	0.042	0.052	–	0.093	0.010
Cr	0.0012	0.0012	–	0.012	0.0002
Ti	0.69	0.75	–	2.65	0.123

dissolving biogenous sediment rich in trace metals. The sediment was originally formed underneath the Equator, then moved northward and downward due to plate motion. This motion brought the calcareous sediment from a zone of accumulation into a zone of dissolution. Thus, one way to concentrate manganese and trace elements is to have organisms precipitate the metals (which they do very efficiently), bring them to the sea floor in shells and fecal matter, and dissolve or oxidize these carriers to obtain a more nearly pure concentrate.

Other mechanisms for concentrating the metals also must exist. For example, the element cobalt tends to be high within ferromanganese accumulating on seamounts under highly oxidizing conditions. Here, presumably, the manganese precipitates extremely slowly out of seawater, at the same time scavenging (co-precipitating catalytically) the chemically similar cobalt.

10.4.4 Ridge Crest Ores: East Pacific Rise. The origin of the ferromanganese deposits on the deep sea floor is hardly in doubt for some of them: those on the crest of the East Pacific Rise. Here seawater circulates through cracks in the newly forming crust, reacting with the hot basalt. Recently, hot vents resulting from this circulation have been studied close-up, through the windows of the reasearch submarine Alvin (Fig. 10.13).

While hydrothermal reactions undoubtedly play a role in producing *metalliferous sediments* on top of the East Pacific Rise, the process of enrichment with various metals other than iron is still poorly understood. Many elemental and isotopic ratios within the deposits are similar to those in seawater and in deposits from non-Ridge areas. Just how much of the material is delivered from the basalt, how much from seawater? One clue to the problem is in the comparison of fresh with altered basalt: whatever is missing from the altered basalt was given off into the seawater. But was it precipitated near the vent? One possibility is that part of the metal enrichment on the East Pacific Rise — especially the very high iron content — is derived from hydrothermal activity, while the rest is scavenged by the freshly formed hydroxide. Scavenging by fresh iron hydroxide is a standard

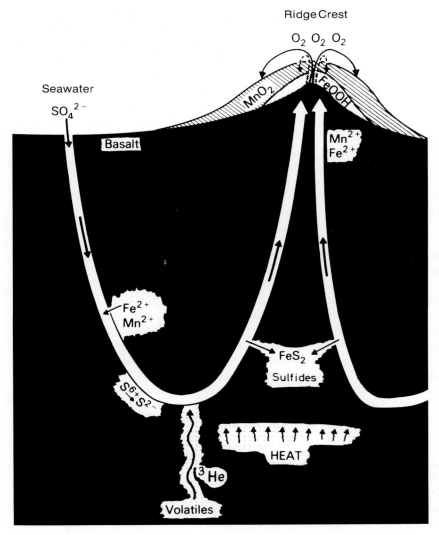

Fig. 10.13. Model of metallogenesis at oceanic spreading centers. Seawater is drawn into the cooling basalt, exits at the crest, in hot vents. Exhaled fluids are highly enriched in iron and manganese, which precipitate near the crest. (From E. Bonatti, 1975, Annu Rev Earth Planet Sci 3: 401)

procedure in chemical experiments, for removing trace metals from solutions at very low concentrations.

10.4.5 Red Sea Ore Deposits. A more specialized case of Ridge Crest accumulation is the heavy metal deposits in the Red Sea, which are of great economic interest. Promising metalliferous deposits occur in the Atlantis II-Deep, named after the Woods Hole research vessel. The basin, which lies

244

offshore of Mecca, was found in 1963 by the British vessel Discoverer; Atlantis II of Woods Hole explored it in 1964 and 1965; the German Meteor was there in 1965; Woods Hole's Chain in 1966; the prospecting vessel Wando River in 1969; and several times since, visits have been made by the Valdivia and the Sonne, economic research vessels from Germany. The Glomar Challenger paid a visit in 1972, drilling in this area (Fig. 10.14).

What is it that attracts all this attention to the Red Sea?

In the central Red Sea — an active spreading center which opened only a few million years ago — there are several enclosed basins, the "deeps". The Atlantis II Deep (Fig. 10.14) is more than 2000 m deep, and only 6 by 15 km in area. The bottom is filled with a hot salt brine with a temperature of about $60°$ C and a salinity of 25%, seven times that of sea water. Iron is 8000 times more concentrated in the brine than in seawater, zinc 500 times, copper 100 times. The sediment below the brine is incredibly colorful, brick-red layers alternating with ocher, white, black, greenish. A variety of minerals provides the coloring; economically the most important are the sulfides in the dark layers. Zinc contents of up to 10%, copper contents of 3% or even 7% were measured. Unfortunately the minerals are extremely fine-grained, which will make extraction difficult.

How did these deposits originate? Apparently two kinds of processes are important. First, we are here in a *volcanic region,* that is, the axis of a spreading center. Basaltic lavas form the basement, and temperatures increase rapidly downward. Hot, metalliferous solutions may develop from reactions of seawater with the fresh basalt. Second, thick *sedimentary deposits* of Tertiary age are nearby, abutting against the newly forming sea floor. These contain several-hundred-meter-thick salt and gypsum layers. Hot water circulating through such sediments can dissolve out metals and salt and hence produce metalliferous brine issuing into the brine pools. The metals precipitate upon colling, and when oxygen is supplied, by mixing with normal seawater. However, such mixing is greatly obstructed by the high density of the brine, and can only occur at the very tops of the brines. Thus, the metals are trapped in the brine.

Various combinations of these processes can be visualized, and the exact mechanism(s) of the brine and metal formation are by no means clear. The layering, presumably, is due to chemical fractionation between the brine pools: the precipitation of metals will occur at different degrees of mixing of metalliferous brine with seawater, depending on the solution chemistry of the metal. Hence each metal will have its own gradient of concentration decrease upward in the water column of a brine pool, different from that of other metals. The exact level, therefore, at which a metalliferous brine overflows into an adjacent basin will determine the ratio between the metals contained. In this fashion, metals can be fractionated from each other, and one or the other metal can be greatly enriched. Small changes in the sill depth of overflow, or in the concentration gradients within the brine, can result in considerable variations in supply of metals — this can conceivably produce much of the observed layering within the sediments.

Just how rich are the deposits? In the Atlantic II Deep alone there are supposedly 3.2 million tons of zinc, 0.8 million tons of copper, 80,000 tons of lead, 4,500 tons of silver, and even 45 tons of gold. Whether the *net worth* of this

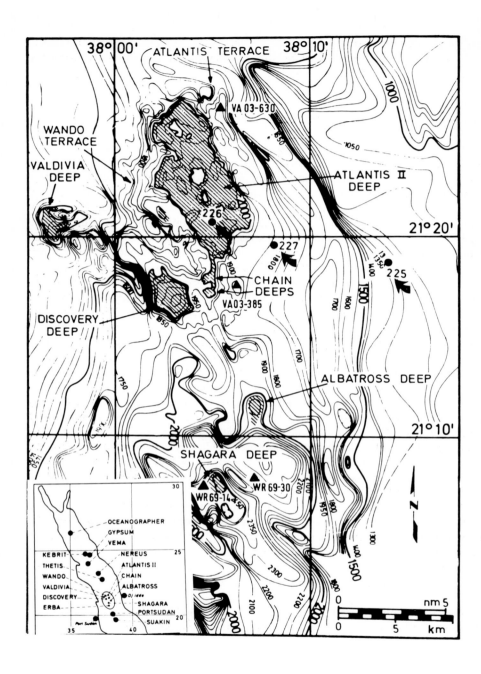

Fig. 10.14. Complicated bathymetry of Atlantis-II-Deep and surrounding deeps in the Central Red Sea. Water depths in meters. Areas covered with brines are *hachured*. *Arrows* point to Glomar Challenger Sites. *Insert* shows concentration of deeps at the median line (spreading center) of the Red Sea. Most of the deeps are named after research vessels

246

ore is significant (after accounting for recovery, processing, and transport) remains to be seen. However, as far as deep sea metals, it is the most promising occurrence found so far.

10.5 Dumping on the Sea Floor

Industrial man uses energy and raw materials and produces waste. The amount of waste produced by the various nations is directly correlated with the gross national product. Most of the waste, such as sewage sludge, is relatively innocuous, especially when introduced in amounts that the ocean can absorb by dilution and bacterial action. Such sludge, in fact, merely acts as fertilizer.

On the other hand, when dumping of sludge proceeds on a grand scale in a limited area, problems can arise. For example off New York, foul bottom conditions were reported from shelf areas in the regions of dumping, with diseased benthic organisms. Fishing activities, especially lobster fishing, are adversely affected under these circumstances. For large coastal cities, however, the value of the sea floor as a dumping ground probably exceeds the value of the seafood from offshore, so that solutions to such conflict are not simple. Where sewage solids settle out in bays close to cities, one potential problem is the stirring up of such material by storms, and the fouling of heavily used recreational areas.

The study of such problems has become an important subject in environmental marine geology; it is a field that is likely to grow.

The fate of radioactive substances on the sea floor may become an extremely important problem in this context. Nuclear power generation is a reality and is spreading quickly. By 1977, western and eastern countries together were operating about 200 nuclear power plants, with new ones being built or planned by an ever-increasing number of nations. Wastes accrue from the fission process (mainly unwanted fission products). The question of the ultimate fate of these wastes is unresolved.

Deep sea dumping is widely used for dangerous chemicals and also for radioactive substances. Such dumping is a classic case of cost externalization. The agency doing the dumping reaps the benefit of being rid of the noxious material, which arose from its economic activities, while the risk is spread evenly to all using the oceans, including — in the case of long-lived radioactive substances — its living organisms and their future generations as well as future generations of mankind.

One important task of marine geologists studying the sea floor — and perhaps the most difficult yet — will be to help assess the risks with which we burden the future.

Epilog

We have attempted, in this brief survey of sea floor studies, to show where we have been and where we are now. We may, perhaps, be permitted a guess as to where we are going.

The recent history of sea floor studies has been characterized by the application of *geophysics:* the grand motions of the continents and the sea floor were discovered by continuous echo sounding (for bathymetry), by the mapping of earthquake foci and of heat flow, by seismic profiling, and especially by the mapping of paleomagnetism.

At the present time, great advances are being made by the application of *geochemistry* to sea floor investigation, in all its manifestations. The oozes and clays recovered by drilling into the deep sea floor contain the geochemical history of the planet which is intertwined with tectonics, climate, ocean circulation, and the evolution of life. The study of carbonates, organic matter, and metalliferous deposits is making great strides, and it is being integrated with the other earth sciences, especially those concerned with the history of the ocean and of climate evolution. Likewise, the evolution of seawater is again becoming the focus of attention: the discovery of intense hydrothermal circulation at the East Pacific Rise has introduced an entirely new spectrum of arguments.

Also, *biology* is getting its turn again, a century after Darwin, to contribute to the earth sciences in a major way. Already, biological studies of the sea floor have yielded discoveries of fundamental importance. One is the great diversity of the starved populations on the deep sea floor, for which environment the traditional diversity-generating mechanisms are difficult to envision. Another is the extraordinary extent to which the chemistry of the ocean is controlled by life processes. And now we have the exciting revelation of an entirely new kind of deep sea community, the hot vent assemblage on the East Pacific Rise (Fig. E.1). How did this community arise? How does it maintain itself, and interact with the normal deep sea biota? These questions are already eliciting a host of new and interesting speculations. One hypothesis proposes that these communities do not depend on sunlight for their energy, through photosynthesis, but that they represent a food chain whose base is chemosynthesis.

We believe it is the *history of life* and the *history of environment,* as recorded in sediments, which will next advance the most from sea floor studies. The problems associated with the history of life and its relationship to the history of ocean chemistry and climate are extremely intricate and challenging. Biostratigraphy — the art of telling time from sequences of fossils — has made enormous progress thanks to long sediment sections from deep sea drilling. It is now at the point, through integration with geophysical and geochemical concepts and methods, where we can catch a glimpse of the rates of change of planktonic and benthic

248

E.1. A new kind of deep sea benthos. Pogonophore-like tube-worms, several feet long, and a host of other creatures crowd around the hot vents of the Galápagos spreading center. "Rose Garden" vent site, Alvin Dive 984. (Photo courtesy R. Hessler, S.I.O.)

population through geologic time, and of the possible reasons for extinctions and radiations.

The study of deep sea sediment sequences has forced geologists to think about stratigraphy on a global scale. The shelf sequences are characterized by fluctuations in sea level, and the endless alternations of transgressions and regressions that go with them. The global signal in these fluctuations, the eustatic changes in sea level, is felt on the deep sea floor as changes in facies distributions produced by changes in the chemistry and geochemistry of the ocean. In the continental margins, especially the passive ones, sea level fluctuations are recorded in the geometry of sediment bodies of vast volumes. The next step, then, is to find the exact history of sea level fluctuations, the amplitudes, periodicities (if any), the rates of change, and especially the causes of sea level change. This will allow a stratigraphic tie-in for the shelf sediment sequences, both exposed and submerged, and the deep sea sequences of the various ocean basins. When this is done, with the aid of biostratigraphy and physicochemical stratigraphy (especially paleomagnetism and stable isotopes), the history of the face of the planet, and the living things on it, will appear in a new light.

List of Books and Symposia

Andel Van TjH, Shor GG (1964) Marine geology of the Gulf of California: a symposium. Am Assoc Pet Geol Mem

Andel Van TjH, Heath GR, Moore TC (1975) Cenozoic tectonics, sedimentation and paleooceanography of the central equatorial Pacific. Geol Soc Am Mem 143

Andersen NR, Malahoff A (eds) (1977) The fate of fossil fuel CO_2 in the oceans. Plenum Press, New York

Ayala-Castanares A, Phleger FB (eds) (1959) Coastal lagoons, a symposium. Univ Nacl Autónoma, Mexico City

Ballard RD, Moore JG (1977) Photographic atlas of the Mid-Atlantic Ridge drift valley. Springer, Berlin Heidelberg New York

Barnes H (ed) (1974) Oceanography and marine biology — an annual review, 12. Allen & Unwin, London

Barnes BSK (ed) (1977) The coastline, a contribution to our understanding of its ecology and physiography in relation to land-use management and the pressures to which it is subject. Wiley-Interscience, London

Bascom W (1964) Waves and beaches — the dynamics of the ocean surface. Doubleday, Garden City (NY)

Bathurst RGC (1975) Carbonate sediments and their diagenesis, 2nd edn Elsevier, Amsterdam

Bentor YK (ed) (1980) Marine phosphorites-geochemistry, occurrence, genesis. Soc Econ Paleontol Mineral Spec Publ 29 Tulsa, Oklahoma, 1–247

Berner RA (1971) Principles of chemical sedimentology. McGraw-Hill New York.

Berner RA (1980) Early diagenesis, Princeton Univ. Press, Princeton/New Jersey

Bischoff JL, Piper DZ (eds) (1979) Marine geology and oceanography of the Pacific Manganese Nodule Province. Plenum Press, New York

Blackett PMS, Bullard E, Runcorn SK (1965) A symposium on continental drift. Philos Trans R Soc London Ser A 258:373

Blatt H, Middleton G, Murray R (1972) Origin of sedimentary rocks. Prentice-Hall, Englewood Cliffs

Borch von der CC (ed) (1978) Synthesis of deep-sea drilling results in the Indian Ocean. Reprinted Mar Geol 26

Bouma AH, Brouwer A (eds) (1964) Turbidites. Elsevier, Amsterdam

Broecker WS (1974) Chemical oceanography. Harcourt Brace Jovanovich, New York

Burk CA, Drake CL (eds) (1974) The geology of continental margins. Springer, Berlin Heidelberg New York

Clark DL, Whitman RR, Morgan KA, Mackey SD (1980) Stratigraphy and glacial-marine sediments of the Amerasian Basin, Central Archiv Ocean. Geol Soc Amer Spec Paper 181, Boulder, Colo

Cline RM, Hays JD (eds) (1976) Investigation of Late Quaternary paleoceanography and paleoclimatology. Geol Soc Am Mem 145

Conchie KC (1976) Plate tectonics and crustal evolution. Pergamon Press, New York

Cook HE, Enos P (eds) (1977) Deep water carbonate environments. Soc Econ Paleontol Mineral Spec Publ 25

Cox A (ed) (1973) Plate tectonics and geomagnetic reversals. Readings with introductions. WH Freeman, San Francisco

Crimes TP, Harper JC (eds) (1970) Trace fossils. Geol J Spec Issue 3

Davis RA (ed) (1978) Coastal sedimentary environments. Springer, Berlin Heidelberg New York

Davis RA, Ethinton RL (eds) (1976) Beach and nearshore sedimentation. Soc. Econ. Paleontol Mineral Spec Publ 24

Degens ET, Ross DA (eds) (1969) Hot brines and recent heavy metal deposits in the Red Sea. Springer, Berlin Heidelberg New York

Degens ET, Ross DA (eds) (1974) The Black Sea — geology, chemistry and biology. Am Assoc Pet Geol

Dott RH, Shaver RH (eds) (1974) Modern and ancient geosynclinal sedimentation. Soc Econ Paleont Mineral Spec Publ 19

Drake CL, Imbrie J, Knauss JA, Turekian KK (1978) Oceanography. Holt Rinehart & Winston, New York

Dunbar CO, Rodgers J (1957) Principles of stratigraphy. Wiley, New York

Ekman S (1953) Zoogeography of the sea. Sidgwick & Jackson, London

Emery KO (1960) The sea off southern California. Wiley, New York

Emiliani C (ed) (1981) The sea vol 7. The Ocean Crust. Wiley-Interscience, New York

Fairbridge RW (ed) (1966) The encyclopedia of oceanography. Reinhold, New York

Fairbridge RW, Bourgeois J (eds) (1978) The encyclopedia of sedimentology. Dowden Hutchinson & Ross, Strondsburg Pa

Fischer AG, Judson S (eds) (1975) Petroleum and global tectonics. Univ Press, Princeton

Flint RF (1971) Glacial and quaternary geology. Wiley, New York

Frey RW (1975) The study of trace fossils. A synthesis of principles, problems, and procedures in ichnology. Springer, Berlin Heidelberg New York

Friedman GM, Sanders JE (1978) Principles of sedimentology. Wiley, New York

Frost SH, Weiss MP, Saunders JB (1977) Reefs and related carbonates — ecology and sedimentology. Studies in geology no 4. Am Assoc Pet. Geol

Funnell BM, Riedel WR (eds) (1971) The micropaleontology of oceans. Univ Press, Cambridge

Gilluly J, Waters AC, Woodford AO (1968) Principles of geology. WH Freeman, San Francisco

Ginsburg RN (ed) (1975) Tidal deposits, a casebook of recent examples and fossil counterparts. Springer, Berlin Heidelberg New York

Glasby GP (ed) (1977) Marine manganese deposits. Elsevier, Amsterdam

Goldberg ED (ed) (1974) The sea, vol 5 (Marine chemistry). Wiley-Interscience, New York

Gorsline DS, Swift DJP (1977) Shelf sediment dynamics: a national overview. Off Int Decade Ocean Explo, NSF, Washington DC

Gribbin J (ed) (1978) Climatic change. Univ Press, Cambridge

Grim RE (1968) Clay mineralogy, 2nd edn. McGraw-Hill, New York

Gross MG (1971) Oceanography, 2nd edn. Merril, Columbus (Ohio)

Hallam A (ed) (1967) Depth indicators in marine sedimentary environments. Marine geology, Spec. Issue, vol 5. Elsevier, Amsterdam

Hallam A (1973) A revolution in the earth sciences. Continental drift to plate tectonics. Clarendon Press, Oxford

Haq BU, Boersma A (1978) Introduction to marine micropaleontology. Elsevier, Amsterdam

Hay WW (ed) (1974) Studies in palaeooceanography. Soc Econ Paleontol Mineral Spec Publ 20

Hays JD (ed) (1970) Geological investigations of the North Pacific. Geol Soc Am Mem 126

Hedgpeth JW (ed) (1957) Treatise on marine ecology and paleoecology. Geol Soc Am Mem 67 (1)

Hedley RH, Adams CG (eds) (1976) Foraminifera. Academic Press, London New York

Heezen BC, Hollister CD (1971) The face of the deep. Oxford Univ Press, New York

Heezen BC, Tharp M (1961) Physiographic diagram of the South Atlantic Ocean. Geol Soc Am, New York

Heezen BC, Tharp M (1964) Physiographic diagram of the Indian Ocean. Geol Soc Am, New York

Heezen BC, Tharp M, Ewing M (1959) The floors of the oceans. I. The North Atlantic. Geol Soc Am Spec Pap 65

Heirtzler JR et al (eds) (1977) Indian Ocean geology and biostratigraphy. Am Geophys Union

Hermann Y (ed) (1974) Marine geology and oceanography of the Arctic seas. Springer, Berlin Heidelberg New York

Hess HH (1962) History of ocean basins. In: Engel AEJ et al (eds) Petrological studies: a volume in honor of A. F. Buddington. Geol Soc Am, New York

Hill MN (ed) (1963) The sea, vol 3. The earth beneath the sea. Wiley, London

Holland HH (1978) The chemistry of the atmosphere and oceans. Wiley, New York

Holmes A (1965) Principles of physical geology, 2nd edn. Ronald Press, New York

Horn DR (ed), (1972) Ferromanganese deposits on the ocean floor. Off Int Decade Ocean Explor, NSF, Washington DC

Hough JL (ed) (1951) Turbidity currents and the transportation of coarse sediments to deep water. Soc Econ Paleontol Mineral Spec Publ 2

251

Hsü KJ, Jenkins H (eds) (1974) Pelagic sediments on land under the sea. Int Assoc Sedimentol Spec Publ 1

Imbrie J, Imbrie KP (1979) Ice ages, solving the mystery. Enslow, Short Hills NJ

Imbrie J, Newell N (eds) (1964) Approaches to paleoecology. Wiley, New York

Ingle JC (1966) The movement of beach sand. Elsevier, Amsterdam

Jones OA, Endean R (eds) (1973–1977) Biology and geology of coral reefs, 4 vols, Academic Press, London

Kaplan IR (ed) (1974) Natural gases in marine sediments. Plenum Press, New York

Kaufman W, Pilkey 0 (1979) The beaches are moving. Anchor Press, Garden City NY

Kent P (ed) (1980) The evolution of passive continental margins in the light of recent deep drilling results. Phil Trans R Soc London A 294; 1–208

King CAM (1972) Beaches and coasts. Arnold, London

Komar PD (1976) Beach processes and sedimentation. Prentice-Hall, Englewood Cliffs

Kuenen PhH (1950) Marine geology. Wiley, New York

Lauff GA (ed) (1967) Estuaries. Am Assoc Adv Sci, Washington DC

LePichon X, Francheteau J, Bonnin J (1973) Plate tectonics. Elsevier, Amsterdam

Lipps JH et al (1979) Foraminiferal ecology and paleoecology. SEPM Short Course no 6. Soc Econ Paleontol. Mineral

Lisitzin AP (1972) Sedimentation in the world ocean. Soc Econ Paleontol Mineral Spec Publ 17

Maxwell AE (ed) (1970) The sea. Ideas and observations in the study of the seas, vol IV. New concepts of sea floor evolution. Wiley-Interscience, New York

McCave IN (1976) The benthic boundary layer. Plenum Press, New York

Menard HW (1964) Marine geology of the Pacific. McGraw-Hill, New York

Mero JL (1965) The mineral resources of the sea. Elsevier, Amsterdam

Middleton GV (ed) (1965) Primary sedimentary structures and their hydrodynamic interpretation. Soc Econ Paleontol Mineral Spec Publ 12

Middleton GV et al (1973) Turbidites and deep-water sedimentation. SEPM Pacific Section Short Course, Anaheim 1973. Soc Econ Paleontol Mineral

Miller RL (ed) (1964) Papers in marine geology (Shepard commemorative volume). Macmillan, New York

Milliman JD (1974) Recent sedimentary carbonates, Part 1: Marine carbonates. Springer, Berlin Heidelberg New York

Millot G (1970) Geology of clays. Springer, Berlin Heidelberg New York

Moore JR (ed) (1971) Oceanography. Readings Sci Am WH Freeman, San Francisco

Morgan JP (ed) (1970) Deltaic sedimentation modern and ancient. Soc Econ Paleontol Mineral Spec Publ 15

Murray J, Renard AF (1891) Deep-sea deposits, based on the specimens collected during the voyage of H.M.S. "Challenger" in the years 1872–1876. "Challenger" Reports. Longmans, London (reprinted by Johnson, London, 1965)

Murray JW (1973) Distribution and ecology of living benthic foraminiferids. Crane Russak, New York

Nairn AEM, Stehli FG (1973) The ocean basins and margins. Plenum Press, New York

National Academy of Sciences (1970) Continental margins — geological and geophysical research needs and problems. Washington DC

Neumann G, Pierson WJ (1966) Principles of physical oceanography. Prentice-Hall, Englewood Cliffs

Payton CE (ed) (1977) Seismic stratigraphy — applications to hydrocarbon exploration. Am Assoc Petr Geol Mem 26

Perkins EJ (ed) (1974) The biology of estuaries and coastal waters. Academic Press, London New York

Phinney RA (ed) (1968) The history of the earth's crust. Univ Press, Princeton

Phleger FB (1960) Ecology and distribution of recent foraminifera. John Hopkins, Baltimore

Pittock AB et al (eds) (1978) Climatic change and variability. Univ Press, Cambridge

Potter PE, Pettijohn FJ (1963) Paleocurrents and basin analysis. Springer, Berlin Heidelberg New York

Press F, Siever R (1978) The Earth, 2nd edn. WH Freeman, San Francisco

Purser BH (ed) (1973) The Persian Gulf — Holocene carbonate sedimentation and diagenesis in a shallow epicontinental sea. Springer, Berlin Heidelberg New York

Ramsay ATS (ed) (1977) Oceanic micropaleontology, 2 vols. Academic Press, London New York

Reading HG (ed) (1978) Sedimentary environments and facies. Blackwell, Oxford

Reineck HE, Singh IB (1973) Depositional sedimentary environments (with reference to terrigenous clastics). Springer, Berlin Heidelberg New York

Riley JP, Chester R (eds) (1976) Treatise on chemical oceanography, vols V, VI, VII. Academic Press, London New York

Riley JP, Skirrow G (eds) (1965) Chemical oceanography, vols I, II. Academic Press, London New York

Saito T, Burckle LH (eds) (1975) Late Neogene Epoch boundaries. Micropaleontology Press, New York

Sarnthein M, Seibold E, Rognon P (eds) (1980) Sahara and surrounding seas. Palaeoecology of Africa vol. 12. Balkema, Rotterdam.

Schäfer W (1972) Ecology and palaeoecology of marine environments. Oliver & Boyd, Edinburg

Schopf TJM (1980) Paleoceanography. Harvard Univ Press, Cambridge (Mass)

Scripps Institution of Oceanography (University of California) (1969) Initial reports of the Deep Sea Drilling Project. Natl Sci Found (US Gov Print Off), Washington DC

Sears M (ed) (1961) Oceanography Am Assoc Adv Sci, Washington DC

Sears M (ed) (1965) Progress in Oceanography, vol IV. The quaternary history of the ocean basins. Pergamon Press, Oxford

Selley RC (1976) An introduction to sedimentology. Academic Press, London New York

Seyfert CK, Sirkin LA (1979) Earth history and plate tectonics. Harper and Row, New York

Shepard FP (1973) Submarine geology, 3rd edn. Harper and Row, New York

Shepard FP, Dill RF (1966) Submarine canyons and other sea valleys. Rand McNally, Chicago

Shepard FP, Phleger FB, van Andel TjH (eds) (1960) Recent sediments, northwest Gulf of Mexico. Am Assoc Petr Geol

Sliter WV, Bé AWH, Berger WH (eds) (1975) Dissolution of deep-sea carbonates. Cushman Found Foraminiferal Res Spec Publ 13

Smith AG, Briden JC (1977) Mesozioc and Cenozoic paleo-continental maps. Univ Press, Cambridge

Stanley DJ (ed) (1969) The new concept of continental margin sedimentation. Am Geol Inst, Washington D

Stanley DJ (ed) (1972) The Mediterranean Sea: a natural sedimentation laboratory. Dowden Hutchinson & Ross, Strondsburg Pa

Straaten van LMJV (ed) (1964) Deltaic and shallow marine deposits. Elsevier, Amsterdam

Sugimura A, Uyeda S (1973) Island arcs: Japan and its environs. Elsevier, Amsterdam

Sullivan W (1974) Continents in motion — the new earth debate. Mc Graw-Hill, New York

Sutton GH, Manghani MM, Moberly R (eds) (1976) The geophysics of the Pacific Ocean basin and its margins. Am Geophys Union Monogr 19

Sverdrup HU, Johnson MW, Fleming RH (1942) The oceans — their physics, chemistry, and general biology. Prentice-Hall, New York

Swift DJP, Palmer HD (eds) (1978) Coastal sedimentation. Benchmark Papers in Geology 142. Dowden Hutchinson Ross, Stroudsburg Pa

Swift DJP, Duane DB, Pilkey OH (1972) Shelf sediment transport: process and pattern. Dowden Hutchinson & Ross, Stroudsburg Pa

Takayanagi Y, Saito T (eds) (1976) Progress in Micropaleontology. Micropaleontology Press, New York

Talwani M, Pitman WC (eds) (1977) Island arcs, deep sea trenches and back-arc basins. Maurice Ewing Ser, vol I. Am Geophys Union, Washington DC

Talwani M, Harrison GG, Hayes DE (eds) (1979a) Deep drilling results in the Atlantic Ocean: ocean crust. Maurice Ewing Ser, vol II. Am Geophys Union, Washington DC

Talwani M, Hay W, Ryan WBF (eds) (1979b) Deep drilling results in the Atlantic Ocean: continental margins and paleoenvironment. Maurice Ewing Ser, vol III. Am Geophys Union, Washington DC

Tarling DH, Runcorn SK (eds) (1973) Implications of continental drift to the earth sciences. Academic Press, London New York

Taylor DL (1977) Proc 3rd Int Coral Reef Symp, vol II. Geology. Univ. Miami, Miami Fla

Tissot B, Welte DH (1978) Petroleum formation and occurrence, a new approach to oil and gas exploration. Springer, Berlin Heidelberg New York

Trask PD (ed) (1939) Recent marine sediments. Am Assoc Pet Geol

Turekian KK (ed) (1971) Late Cenozoic glacial ages. Yale Univ Press, New Haven (Conn)

Turekian KK (1976) Oceans, 2nd edn. Prentice-Hall, Englewood Cliffs
Uyeda S (1978) The new view of the earth — moving continents and moving oceans. WH Freeman, San Francisco
Valentine JW (1973) Evolutionary paleoecology of the marine biosphere. Prentice-Hall, Englewood Cliffs
Watkins JS, Montadert L, Dickerson PW (eds) (1979) Geological and geophysical investigations of continental margins. Am Assoc Pet Geol Mem 29
Wegener A (1966) The origin of continents and oceans. (Translation of the 4th ed. of "Die Entstehung der Kontinente und Ozeane", 1929). Dover Publications, New York
Whitaker JHM (ed) (1976) Submarine canyons and deep-sea fans, modern and ancient. Benchmark Pap Geol 24
Wiens HJ (1962) Atoll environment and ecology. Yale Univ Press, New Haven (Conn)
Wilson JT (ed) (1972) Continents adrift. Readings Sci Am. WH Freeman, San Francisco
Wilson JT (ed) (1976) Continents adrift and continents aground. Readings Sci Am. WH Freeman, San Francisco
Wooster WS (ed) (1970) Scientific exploration of the South Pacific. Nat Acad Sci Washington DC
Wright HE, Frey DG (eds) (1965) The quaternary of the United States. Univ Press, Princeton
Wright JB (ed) (1978) Mineral deposits, continental drift, and plate tectonics. Benchmark Pap Geol 144
Wyllie PJ (1971) The dynamic earth. John Wiley, New York
Yarborough H et al (1977) Geology of continental margins. AAPG Continuing Education Course Note Series no. 5. Am Assoc Pet Geol
Zenger DH, Dunham JB, Ethington RL (eds) (1980) Concepts and models of dolomitization. Soc. Econ Paleontologists and Mineralogists. Special Publ. Nr. 28, 1–320. Tulsa Oklahoma.
Zenkovitch VP (1967) Processes of coastal development. Oliver & Boyd, Edinburgh

Recent Additions

Berger WH, Bé AWH, Vincent E (eds) (1981) Oxygen and carbon isotopes in foraminifera. Paleogr climat ecol 33
Blanchet R, Montaden L (eds) (1981) Geology of Continental margins. Oceanol Acta 4, suppl
Burns RG, Burns VM (eds) (1979) Marine minerals. Mineralog Soc Amer Short Course Notes 6
LePichon X, Debyser J, Vine F (eds) (1981) Geology of oceans. Oceanol Acta 4, suppl
Riedel WR, Saito T (eds) (1979) Marine plankton and sediments. Micropaleont Spec Publ 3
Rona PA, Lowell RP (eds) (1980) Seafloor spreading centers: Hydrothermal systems. Academic Press, New York
Shepard FP, Marshall NF, McLoughlin PA, Sullivan GG (1979) Currents in submarine canyons and other sea valleys. AAPG Studies in Geology 8

Appendix

A1 Conversions Between Common US Units and Metric Units

Temperature

$C = (F\text{-}32) \cdot 5/9$

$F = C \cdot 9/5 + 32$ where C is degrees Celsius or "centigrade" and F is degrees Fahrenheit

Freezing point of water: 0° C, 32° F

Boiling point of water: 100° C, 212° F

Typical room temperature: 20° C, 68° F

Degrees Kelvin (=absolute scale)=Celsius+273.2°

Length

1 cm =0.394 inch

1 m =3.281 ft

1 km=0.621 miles

1 cm =10 mm (millimeter)

1 inch=2.54 cm

1 ft=0.305 m

1 mile (statute)=1.609 km

1 mm=1000 µm (micrometer, microns)

Volume

1 liter=1000 milliliters=0.264 US gallons

1 US gallon=3.785 l

1 barrel (oil)=42 gallons

Mass

1 kilogram=1000 grams=2.205 pounds

1 pound=0.454 kilograms

1 metric ton=1000 kg=1.102 short tons

A2   Topographic Statistics

Earth
Equatorial radius = 6378 km
Polar radius = 6356 km
Area = 510×10^6 km^2
Volume = 1.083×10^{12} km^3
Northern hemisphere: 61% oceans
Southern hemisphere: 81% oceans
Sea floor = 71% of Earth's surface

Oceans
Atlantic Ocean (incl. Arctic and marginal seas)
    Area = 107 million km^2
    Volume = 351 million km^3
Indian Ocean
    Area = 74 million km^2
    Volume = 285 million km^3
Pacific Ocean
    Area = 181 million km^2
    Volume = 714 million km^3
Total ocean
    Area = 362 million km^2
    Volume = 1347 million km^3
Depths: for depths see Table 2.2

Source for A2, H. U. Sverdrup et al. (1942) The oceans, Prentice Hall, Engle wood Cliffs, New Jersey.

Shelves: for shelf areas see Table 2.1

A3   The Geologic Time Scale

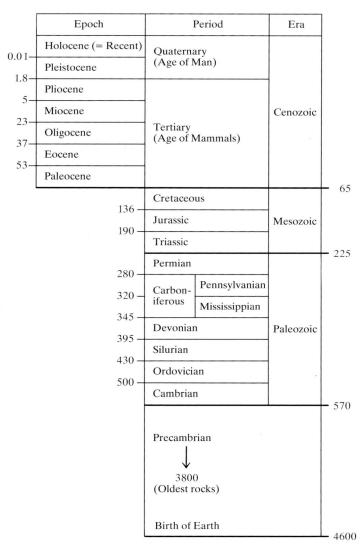

Epoch	Period		Era
Holocene (= Recent)	Quaternary (Age of Man)		Cenozoic
Pleistocene			
Pliocene	Tertiary (Age of Mammals)		
Miocene			
Oligocene			
Eocene			
Paleocene			
	Cretaceous		Mesozoic
	Jurassic		
	Triassic		
	Permian		Paleozoic
	Carbon-iferous	Pennsylvanian	
		Mississippian	
	Devonian		
	Silurian		
	Ordovician		
	Cambrian		

0.01
1.8
5
23
37
53
65
136
190
225
280
320
345
395
430
500
570

Precambrian

↓

3800
(Oldest rocks)

Birth of Earth

4600

Numbers indicate boundary ages in millions of years. Note that the age of fossils, the *Phanerozoic* is only one eighths of Earth's history. All other time belongs to the *Precambrian*. The deep-sea record begins in the Jurassic (see Chap. 9). The Holocene is the youngest and the shortest epoch, it begins with the disappearance of the large ice caps from Canada and Scandinavia, 10,000 years ago.

Sources for A3, any geology text.

257

A4 Periodic Table of the Elements

METALLIC ELEMENTS   NON-METALLIC ELEMENTS

ALKALI METALS ▸	ALKALINE EARTH ELEM. ▸							TRANSITION ELEMENTS									INERT GASES ▸
H 1, 1 Hydrogen																	He 2, 4 Helium
Li 3, 7 Lithium	Be 4, 9 Beryllium											B 5, 11 Boron	C 6, 12 Carbon	N 7, 14 Nitrogen	O 8, 16 Oxygen	F 9, 19 Fluorine	Ne 10, 20 Neon
Na 11, 23 Sodium	Mg 12, 24 Magnesium											Al 13, 27 Aluminium	Si 14, 28 Silicon	P 15, 31 Phosphorus	S 16, 32 Sulfur	Cl 17, 35 Chlorine	Ar 18, 40 Argon
K 19, 39 Potassium	Ca 20, 40 Calcium	Sc 21, 50 Scandium	Ti 22, 48 Titanium	V 23, 51 Vanadium	Cr 24, 52 Chromium	Mn 25, 55 Manganese	Fe 26, 56 Iron	Co 27, 59 Cobalt	Ni 28, 59 Nickel	Cu 29, 64 Copper	Zn 30, 65 Zinc	Ga 31, 70 Gallium	Ge 32, 73 Germanium	As 33, 75 Arsenic	Se 34, 79 Selenium	Br 35, 80 Bromine	Kr 36, 84 Krypton
Rb 37, 85 Rubidium	Sr 38, 88 Strontium	Y 39, 89 Yttrium	Zr 40, 91 Zirconium	Nb 41, 93 Niobium	Mo 42, 96 Molybdenum	Tc 43, 99 Technetium	Ru 44, 101 Ruthenium	Rh 45, 103 Rhodium	Pd 46, 108 Palladium	Ag 47, 108 Silver	Cd 48, 112 Cadmium	In 49, 115 Indium	Sn 50, 119 Tin	Sb 51, 122 Antimony	Te 52, 128 Tellurium	I 53, 127 Iodine	Xe 54, 131 Xenon
Cs 55, 133 Cesium	Ba 56, 137 Barium	Rare* Earths 57–71	Hf 72, 178 Hafnium	Ta 73, 181 Tantalum	W 74, 184 Tungsten	Re 75, 186 Rhenium	Os 76, 190 Osmium	Ir 77, 192 Iridium	Pt 78, 195 Platinum	Au 79, 197 Gold	Hg 80, 201 Mercury	Tl 81, 204 Thallium	Pb 82, 207 Lead	Bi 83, 209 Bismuth	Po 84, 210 Polonium	At 85, 210 Astatine	Rn 86, 222 Radon
Fr 87, 223 Francium	Ra 88, 226 Radium	Ac 89, 227 Actinium	Th 90, 232 Thorium	Pa 91, 231 Protactinium	U 92, 238 Uranium												

ALKALI METALS

ALKALINE EARTH ELEM.

HALOGENS ▸

a   b   c   d

* Rare Earth elements: Series
beginning with Lanthanium 57/139

The table shows symbol, name, atomic number (upper right), and atomic weight of most common isotope (lower right) in each box. The most abundant elements in the Earth's crust are printed in *bold* letters. Certain oceanographic groups of elements may be distinguished as follows (see graph):

a   major cations in seawater (and major elements in salt deposits, and also in marine carbonates)

b   trace elements typically concentrated in ferromanganese concretions on the deep sea floor (Mn-Zn)

c   elements involved in intense biological cycling (C, N, P, Si)

d   elements of the major anions in seawater (also salt deposits) (O, S, Cl, Br) Carbon also belongs into this group (HCO_3^- ion)

Sources for A4, see any chemistry text

A5   Common Minerals

Silicate Minerals
Most abundant, crust-forming minerals. Basic building blocks: SiO_4-tetrahedra (see graph).

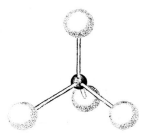

Fig. A5. SiO_4-tetrahedron, a silicon atom surrounded by four oxygen atoms

Quartz SiO_2. Framework structure: each tetrahedron shares all oxygens with other tetrahedra. Found in granitic rocks and in sediments. Opal=disordered hydrated silica ($SiO_2 \cdot nH_2O$)

Feldspars. Framework structure as in quartz. Replacement of Si by Al in some tetrahedra results in negative charge which is balanced by cations. Found in virtually all rocks. Examples:

Albite   Na Al Si_3O_8
Anorthite   Ca $Al_2Si_2O_8$
Microcline   K Al Si_3O_8

Micas. Sheet structure: each tetrahedron is tied to others by three oxygens. Sheets are bonded by cations. Ubiquitous. Examples:

Muscovite   K Al_2(Al Si_3)O_{10}(OH)$_2$
Biotite   K Mg_3(Al Si_3)O_{10}(OH)$_2$ (also usually contains iron)

Clay minerals. Similar to mica, but deficient in cations. See Fig. 8.4 for clay mineral structure. Abundant in sediments and in lowgrade metamorphic rocks. Produced during chemical weathering of rocks with feldspar and mica. Examples:

Montmorillonite (=smectite) (Al,Fe^{3+},Mg)$_3$ (OH)$_2$[(Si,$_{11:1}$Al)$_4O_{10}$]Na $\cdot$ nH_2O
Illite   (K,H_3O)Al_2(H_2O,OH)$_2$[Al Si_3O_{10}]
Chlorite   (Al,Mg,Fe)$_3$(OH)$_2$(Al,Si)$_4O_{10}Mg_3$(OH)$_6$
Kaolinite Al_4(OH)$_8$[Si_4O_{10}]

Hornblende-type minerals. Double chain structure. Abundant in metamorphic rocks, and in many igneous rocks. Example:

Amphibole   (Ca_2Mg_5)[Si_8O_{22}](OH)$_2$

Augite-type minerals. Single chain structure. Abundant in basalts. Example:

Pyroxene   (Mg,Fe)Si O_3

Olivine series. (Mg,Fe)Si O_4 Isolated tetrahedra bonded via cations. Abundant in basalts.

Zeolites. Feldspar-like structure but with large interstices containing water. Common in volcanogenic sediments and in deep-sea clays. Examples:

Phillipsite   (1/2Ca, Na,K)$_3$[$Al_3Si_5O_{16}$]$\cdot$6H_2O
Clinoptilotite   (CaNa_2)[$Al_2Si_7O_{18}$]$\cdot$6H_2O

259

Nonsilicate Minerals

Carbonates. Bulk of biogenous sediments. Examples:

Calcite	$Ca\,C\,O_3$	shells and skeletons, see Table 3.3
Aragonite	$Ca\,C\,O_3$	ditto
Dolomite	$Ca\,Mg\,(CO_3)_2$	diagenesis, see Sect. 3.

Evaporite minerals. Precipitated from seawater. Examples:

Anhydrite	$Ca\,S\,O_4$
Gypsum	$Ca\,SO_4 \cdot 2H_2O$
Halite	$Na\,Cl$
Epsomite	$Mg\,S\,O_4 \cdot 7H_2O$
Bischofite	$Mg\,Cl_2 \cdot 6H_2O$
Carnallite	$K\,Mg\,Cl_3 \cdot 6H_2O$

Iron oxides and sulfides. Ubiquitous in igneous and sedimentary rocks. Examples:

Goethite	$\alpha\text{-FeOOH}$
Lepidocrocite	$\gamma\text{-FeOOH}$
Limonite	$FeOOH \cdot nH_2O$
Hematite	$\alpha\text{-Fe}_2O_3$
Magnetite	Fe_3O_4
Pyrite	FeS_2
Marcasite	FeS_2
Hydrotroilite	$FeS_2 \cdot nH_2O$

Heavy minerals (see Sect. 3.5.2)

Apatite	$Ca_5(PO_4)_3(F,Cl,OH)$
Augite	Ca,Mg,Fe – silicate
Barite	$BaSO_4$
Epidote	Ca,Fe,Al – silicate
Garnet	Mg,Ca,Fe,Al – silicates
Glauconite	K,Fe – mica
Hematite	iron oxide
Hornblende	Ca,Fe,Na,K,Mg,Al – silicates
Ilmenite	$FeTiO_3$
Limonite	iron (hydr)oxide
Marcasite	iron sulfide
Magnetite	iron oxide
Olivine	Mg,Fe,Ca – silicates
Pyrite	iron sulfide
Rutile	TiO_2
Titanite	Ca,Ti – silicate
Zircon	Zr – silicate

Source for A5, any mineralogy textbook

260

A6   Common Rock Types

Igneous rocks   Mineral assemblages (dominated by silicates) which crystallized from hot, fluid magma. Slow cooling produces coarse crystals, typical for intrusive rocks. Fast cooling produces small crystals, typical for extrusive rocks (volcanic lava). Classification is by mineral types (from high-silica to low-silica minerals) and by crystal size.

Intrusive rocks	*Extrusive rocks*	
(coarse crystals)	(fine-grained)	decreasing
granite	rhyolite	proportion
granodiorite	dacite	of SiO_2, increase
diorite	andesite[1]	in Mg and Fe
gabbro	basalt[2]	
peridotite	basalt[2]	↓
dunite	basalt	

[1] typical volcanic rock in island arcs and landward of trenches (Andes!). Partial melting of downgoing oceanic crust appears to be an important process in generating andesite (see Fig. A6.1).
[2] typical basement rock for deep sea floor, that is, ocean crust below sediments.

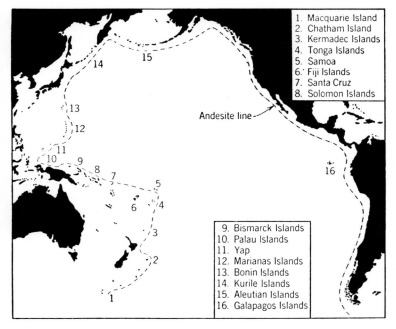

Fig. A6.1 Andesite line. The "andesite line" separates andesitic volcanoes from basaltic ones. The recognition of this boundary predates plate tectonics by many decades (see Fig 1.11). Graph from Kuenen Ph H (1950) Marine Geology. John Wiley and Sons, New York

Sedimentary Rocks. Consolidated sediments. Particle assemblages consisting of minerals, rock fragments, shell, and amorphous substances in various proportions. Produced by weathering, transportation, deposition, alteration, and cementation. Classification is by grain-size and composition. (See Chapter 3.)

Detrital or Clastic Rocks.
(unconsolidated) (consolidated)

gravel	conglomerate	decreasing
sand	sandstone	grain-size
silt	siltstone	
clay	claystone	↓
mud[1]	mudstone	
	shale[2]	

[1] *Mud* is a mixture of silt and clay
[2] *Shale* is the most common sedimentary rock, it is siltsone, claystone, and mudstone which is laminated and parts along bedding planes

Examples of common sand- and siltstones are:
Quartz sandstone: produced through recycling of sedimentary rocks (quartz is resistant to weathering)
Arkose: feldspar-rich sandstone, produced from erosion of granite and granodiorite
Graywacke: cemented "dirty" sand, that is, with abundant clay and rock fragments. Composition is similar to that of sediments accumulating on active margins (see Fig. 2.6).

Chemical Rocks. Assemblages of minerals produced by precipitation from watery solution, either by organisms or inorganically. Classification is by origin and composition.
(biogenous)   (nonbiogenous)
reef-limestone   dolomite
bedded limestone   evaporite
chert[1]
coal
phosphorite

[1] *Chert* denotes a series of rocks ranging from silicified mudstones to very fine-grained quartz rock. The origin of the silica (quartz) is biogenous opal (at least in Mesozoic and younger rocks).

Metamorphic Rocks. Mineral assemblages produced by deformation and recrystallization of igneous rocks or sedimentary rocks through elevated temperature and pressure, with or without addition of material through injection of solutions. Classification is by degree of foliation, types of minerals present, and grain size. Metamorphic rocks occur at active margins (see Fig. 2.6).

(Nonfoliated)
Hornfels (from shale, tuff, or lava)
Marble (from limestone or dolomite)
Quartzite (from quartz sandstone)
Amphibolite (from basaltic material or corresp. sediments)
Granulite (from shale, graywacke, or andesite)

(Foliated)
Slate (from shale, tuff)
Schist (from basalt, andesite, tuff, or shale)
Gneiss (from granite, diorite, schist, shale, etc)

Source for A6, any geology text-book

262

A7 Geochemical Statistics

Chemical composition of crust, ocean, and atmosphere

	Cont. crust	Ocean crust	Seawater	Atmosphere
O	46.3	43.6	85.8	21.0
Si	28.1	23.9	–	–
Al	8.2	8.8	–	–
Fe	5.6	8.6	–	–
Ca	4.2	6.7	0.04	–
Na	2.4	1.9	1.1	–
Mg	2.3	4.5	0.14	–
K	2.1	0.8	0.04	–
Ti	0.6	0.9	–	–
H	0.14	0.2	10.7	tr
P	0.10	0.14	–	–
Cl	–	–	2.0	–
N	–	–	–	78.1
Ar	–	–	–	0.9
CO_2	–	–	tr	0.03
	100.0	100.0	99.8	100.0

The crust is mainly made of alumino-silicates, with the alkaline and earth-alkaline elements as cations, and with iron inside and outside of the silicate minerals. Continental crust has a composition resembling that of granodiorite, ocean crust that of basalt. Seawater is largely a solution of sodium chloride. The atmosphere is a mixture of nitrogen and oxygen, both intimately tied to biological cycling.

Composition of Igneous and Sedimentary Rocks

	Igneous (Continent)	Sedimentary (Continent)	Tholeiitic basalt (deep sea)	Alkali olivine basalt (ocean volcanoes)	Red Clay (deep sea)
SiO_2	59.1	57.9	50.2	48.2	53.7
Al_2O_3	15.3	13.3	16.2	16.5	17.4
FeO	3.8	2.1	7.1	7.6	0.5
Fe_2O_3	3.1	3.5	2.6	4.2	8.5
CaO	5.1	5.9	11.4	9.1	1.6
Na_2O	3.8	1.1	2.8	3.7	1.3
MgO	3.5	2.7	7.7	5.3	4.6
K_2O	3.1	2.9	0.2	1.9	3.7
H_2O	1.1	3.2	$<1^a$	$<1^a$	6.3
TiO_2	1.1	0.6	1.5	2.9	1.0
MnO	–	0.1	0.2	0.2	0.8
P_2O_5	0.3	0.1	0.1	0.5	0.1
CO_2	0.1	5.4	–	–	0.4
C (org)	–	0.7	–	–	0.1
SO_3	–	0.5	–	–	–
	99.4	100.0	99.8	99.9	100.0

[a] assumed as zero in the calculation

On the whole, igneous rocks and the sediments derived from them (mainly shales) have very similar composition (columns 1 and 2). Note the depletion, in the sediments, of sodium (Na_2O) much of which ends up in sea water. Note also the high CO_2 content in sediments (carbonates), the high carbon content (coaly matter) and the sulfur (evaporites). The sea floor basalts differ from continental igneous rocks in the abundance and oxidation state of iron, in the calcium content, in magnesium, and especially in potassium. Red clay has a composition much like sedimentary rock in general, except for depletion in calcium carbonate (dissolution!), and enrichment in manganese. Also note the difference in sulfur (no evaporites) and in carbon (high degree of oxidation). (Sources for A7: Fairbridge R. W. ed. 1972 The encyclopedia of geochemistry and environmental sciences, Van Nostrand Reinhold Co. New York; A. E. J. Engel and C. G. Engel, 1971 in A. Maxwell [ed] The Sea vol. 4 pt. 1, 465–519, John Wiley and Sons, New York).

A8   Radio-Isotopes and Dating

General
Certain atoms have a nucleus which emits radiation, whereupon the nature of the atom changes. Atoms of the same kind are called *isotopes* and those which emit radiation from the nucleus are *radioactive* isotopes. Many elements have both *stable* and *unstable* (= radioactive) *nuclides* (= isotopes), but elements heavier than lead (see A4) generally tend to be unstable. The radiation emitted from a nucleus may be α, β, or γ-radiation. An alpha particle consists of two neutrons and two protons (= helium nucleus); a beta particle is a high-speed electron, and gamma rays are like x-rays except more energetic.

The probability that a given atomic nucleus emits radiation (or "decays") is independent of its age and of environment (pressure, temperature, chemical conditions). From this it follows that the number of atoms in a given set of radioactive isotopes decreases according to a simple exponential law, which may be written:

$$N / N_o = e^{-\lambda t}$$

where N / N_o is the proportion of remaining atoms after time t, and λ is the *decay constant*. Solving the equation for $N/N_o = 1/2$ we obtain a value for t which is called the *half-life*

$$t_{1/2} = \ln 2 / \lambda$$

The decaying idotope is called *parent* and the resulting new isotope is called *daughter*. Geochronological dating is done by assessing the abundance of parent and/or daughter elements. When N/N_o ist known, and λ, then t can be calculated.

Types of Radioactivity
Three kinds of radioactivity are useful in measuring time in the geological record.

The first is *primary* radioactivity from radioisotopes with very long half-lives, which have been around since the Earth was formed. These isotopes are potassium-40, rubidium-87, thorium-232,

264

uranium-235, and uranium-238 (^{40}K, ^{87}Rb, ^{232}Th, ^{235}U, ^{238}U). The decay of ^{40}K to ^{40}Ar has been used to date volcanic rocks *(potassium-argon dates)*. Such dating delivered the time scale for magnetic lineations on the sea floor (Chap. 1). The decay of ^{87}Rb to ^{87}Sr has been used to date igneous rocks and meteorites, and to establish the timing of metamorphism *(rubidium-strontium method)*. The method is also used to study inhomogeneities in the upper mantle, as reflected in different ^{87}Sr/^{86}Sr ratios in oceanic basalts on the seafloor. ^{232}Th decays to ^{208}Pb in a series involving six alpha steps (each decreasing the atomic weight of the preceding parent by 4). ^{208}Ti also is produced in this decay. The ^{232}Th decay has been used to date the time of crystallization of certain minerals. The two *uranium decay series* are especially important in dating Pleistocene deposits (see also Sect. 9.3.5).

Another kind of radioactivity is called *secondary;* it consists in the alpha, beta, and gamma emissions from the daughter elements along a decay series. The third kind of radioactivity is *cosmic-ray induced*. Certain radioisotopes are continually being produced by cosmic ray bombardment of the atmosphere and the ocean. Their abundance represents an equilibrium between new production on Earth, and the decay by radioactivity. Examples are carbon-14, hydrogen-3 (=tritium), beryllium-7, beryllium-10, and silicon-32.

A fourth kind of radioactivity — the man-made type — is useful for measuring the rate of rapid geologic processes, such as water movments, dispersion of sediments, sediment mixing on the sea floor, and growth rate of skeleton-producing organisms. The degree to which bomb-produced tritium and carbon-14 are dispersed in the ocean, for example, contains clues about rates of deep water formation and rates of carbon cycling. The depth to which plutonium penetrates sediment on the sea floor tells somethin about the activity of benthic organisms. Some short-lived isotopes in the natural uranium decay series also are used for the study of such processes.

Uranium Decay Series

There are two such series, one for ^{238}U and one for ^{235}U. the heavier uranium is the more abundant of the two isotopes (99.27% versus 0.72%). The two series proceed by alpha and beta decay, as follows (half-life below each arrow):

$$^{238}U \xrightarrow[4.49 \cdot 10^9 \text{ yrs}]{\alpha} {}^{234}Th \xrightarrow[24.1 \text{ ds}]{\beta} {}^{234}Pa \xrightarrow[1.18 \text{ min}]{} {}^{234}U \xrightarrow[2 \cdot 48 \cdot 10^5 \text{ yrs}]{\alpha} {}^{230}Th \xrightarrow[7.5 \cdot 10^4 \text{yrs}]{\alpha} {}^{226}Ra$$

$$^{226}Ra \xrightarrow[1622 \text{ yrs}]{\alpha,\alpha,\alpha,\beta,\beta,\alpha} {}^{210}Pb \xrightarrow[22 \text{ yrs}]{\beta,\beta,\alpha} {}^{206}Pb \text{ (stable)}$$

$$^{235}U \xrightarrow[7.13 \cdot 10^8 \text{ yrs}]{\alpha} {}^{231}Th \xrightarrow[25.6 \text{ hrs}]{} {}^{231}Pa \xrightarrow[3.25 \cdot 10^4 \text{ yrs}]{\alpha} {}^{227}Ac \xrightarrow[22 \text{ yrs}]{\beta} {}^{227}Th \xrightarrow[18.6 \text{ ds}]{\alpha} {}^{223}Ra$$

$$^{223}Ra \xrightarrow[11.1 \text{ ds}]{\alpha,\alpha,\alpha,\beta,\alpha,\beta} {}^{207}Pb \text{ (stable)}$$

The ratios between start and end elements, and the ratio between the stable end products *(lead-lead method)* have been used for geochronometry on long time scales. The relatively short-lived intermediate products offer opportunities for dating on time scales of the Pleistocene. Both sediment accumulation and the age of raised coral terraces have been dated by uranium series analysis.

Sediment dating proceeds from the observation that uranium is much more soluble than thorium (Th) or protactinium (Pa), daughter products in the decay series. The decay products thorium and protactinium enter the sediment at the stages shown in Fig. A8.1. At this point the particular elements form insoluble compounds and they also exist long enough to reach the sea floor as precipitates within particles. Decay then continues within the sediment at a rather slow rate: ^{230}Th has a half-life of 75,000 years, and ^{231}Pa one of 32,500 years.

From Fig. A8.1 it is obvious that we should find, within the sediment, the thorium-230 and the protactinium-231 in *excess* of what is expected to be delivered by the uranium-238 and the uranium-235 present in the sediment. Also, since these daughter isotopes decay slowly in their turn (to radium-226 and actinium-227, respectively), this excess abundance should decrease with the age of the sediment, that is, downward in a core. These expectations are fulfilled, and the decreasing abundance of excess thorium-230 (or of excess protactinium-231) downcore yields a measure for the sedimentation rate (Fig. A8.2).

In the figure, measurements on thorium activity are plotted against depth in core. Thorium activity is shown as the difference (in disintegrations per minute) between the activity of ^{230}Th and of

265

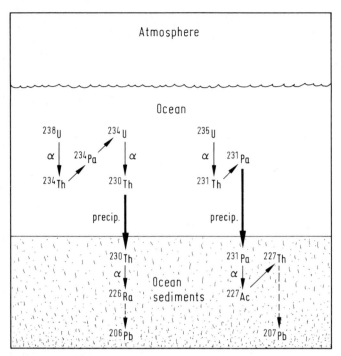

Fig. A8.1. Behavior of uranium series isotopes in the ocean. Uranium is much more soluble than thorium and protactinium. These elements, therefore, precipitate when formed by decay of their immediate parent. ^{230}Th and ^{231}Pa are the first long-lived isotopes in their respective series. Thus, they have time to reach the sediment. (From W. M. Sackett, 1964, Ann N Y Acad Sci *119:*340)

^{234}U, its immediate parent. This difference is the *unsupported* or *excess* thorium-230. It is assumed to be the remnant of thorium originally entering the sediment at its former surface. The broken line shows where one would expect to plot the measurements if the original activity of excess thorium had been 10 dpm, and the sedimentation rate had been unchanged at 2.4 cm/1000 years. The fit is reasonable but not perfect: the thorium activity is too high near the top of the core and too low between 2 and 4 m depth-in-core.

Carbon-14 Dating (see also Sect. 6.7.1)

Carbon-14 arises in the atmosphere from the interaction of nitrogen-14 (the most abundant isotope in the air) with slow moving neutrons which owe their origin to cosmic ray bombardment of the atmosphere. About 10 kg of carbon-14 is produced per year, in this fashion. The half-life of ^{14}C is 5700 years. Thus, the proportion of radiocarbon which disintegrates each year (back to nitrogen-14) is one hundredth of 1.2% of the total radiocarbon present. For steady state, the amount formed equals the amount decaying, and the total mass of radiocarbon present, therefore, should be near 80,000 kg. Each living organism (and each carbonate shell abuilding) has its share of this radiocarbon within it. Bomb-produced ^{14}C has considerably increased the total amount (and the share).

For dating purposes the condition before the introduction of man-made radiocarbon is taken as the normal state (A.D. 1950). In reality there is no steady state, as was shown by tree-ring dating for the last 8,000 years. Radiocarbon dates may deviate from true ages by as much as 10% in this period (and presumably for earlier periods as well).

The maximum range of the radiocarbon method in sediments is about 40,000 yrs, which corresponds to seven half-lives or 1/128 of the original activity. One measures the ratio between radiocarbon and the total carbon present. The original ratio of radiocarbon to stable carbon

266

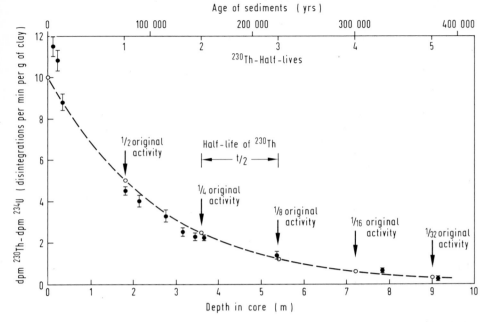

Fig. A8.2. Thorium-dating of deep-sea core V 12-122 raised from the Caribbean deep sea floor (17°00′ N, 74°24′ W, 2800 m, core length 10.95m). The part of thorium-230 activity which is unsupported by uranium-234 decay in the sediment is plotted as disintegrations per minute, on a carbonate-free basis. The dashed line and hollow circles show the decrease of activity expected if the initial value is 10dpm, and the rate of sedimentation is 2.4 cm/1000 yrs, and assuming no variation of the supply of ^{230}Th through time. (Data from T. L. Ku and W. S. Broecker, 1966, Science 151:448)

(carbon-12 and a minor amount of carbon-13) is about one in 10^{12} (one in one thousand billion). The age is calculated from the equation

$$(^{14}C/\text{total C})_{\text{present}} = (^{14}C/\text{total C})_{\text{initial}} \cdot e^{-\lambda t}$$

where the decay constant is 1/8200.

From this follows

$$t = 8200 \ln \frac{(^{14}C/C)_{\text{initial}}}{(^{14}C/C)_{\text{present}}}$$

The initial ratio is assumed to be that of living matter in AD 1950. There also is a correction for differential incorporation of ^{14}C and ^{12}C into various types of living and skeletal matter. This correction is based on the measurement of $^{13}C/^{12}C$ ratios.

Either carbonate or organic carbon may be dated. In sediments, contamination from redeposition can be a problem in fine-grained carbonate, as well as in organic carbon, especially near continents. Also, the ^{14}C content of organic carbon can change through exchange with the atmosphere, during improper storage of sediment. Sediment mixing is an important problem in interpreting ^{14}C dates of marine deposits (Sect. 6.7.1). This problem does not arise in the dating of single large shells, of course.

The amount of carbon necessary for routine dating is 10 grams of carbon (or 100 g of carbonate). Refined methods can cope with amounts much smaller than this.

A9 Systematic Overview for Major Groups of Common Marine Organisms Important in Sea-Floor Processes

The number of species of living organisms on Earth is not known, estimates range from 3 to 10 million. As far as animals (including protozoans, i.e., eukaryotic unicellular organisms without chlorophyll), there are well over a million species, of which 16% (160,000) live in the ocean. On land, some 75% of the animal species are insects. Hence, the number of *noninsect* species in the sea is two thirds of the total noninsect species. Benthic species outnumber pelagic ones by 50:1.

On the basis of similarity (and presumed common origin, therefore) species are grouped into *genera* (sing. *genus*), genera are grouped into *families*, these into *orders*, these into *classes*, and classes finally into *phyla* (sing. *phylum*) (see Table A9.1).

Table A9.1. Taxonomic classification: four examples. (J. L. Sumich, 1976, An introduction to the biology of marine life. WC Brown Co., Dubuque, Iowa)

Taxonomic Category	Blue Whale	Common Dolphin	Purple Sea Urchin	Giant Kelp
Kingdom	Animalia	Animalia	Animalia	Plantae
Phylum/ Division	Chordata	Chordata	Echinodermata	Phaeophyta
Class	Mammalia	Mammalia	Echinoidea	Phaeophycae
Order	Mysticeti	Odontoceti	Echinoida	Laminariales
Family	Balaenopteridae	Delphinidae	Strongylocentrotidae	Lessoniaceae
Genus	*Balaenoptera*	*Delphinus*	*Strongylocentrotus*	*Macrocystis*
Species	*musculus*	*delphis*	*purpuratus*	*pyrifera*

The highest taxonomic category is the *kingdom*. Traditionally two kingdoms have been recognized, the "animals" and the "plants", a division which goes back to the time of the 18th century naturalist C. Linné. The biologist E. Haeckel (in 1866) recognized the *protists* (unicellular organisms) as a separate kingdom, and H. F. Copeland (in 1938) elevated the *monera* (Bacteria and blue-green algae) to this rank. A five-kingdom system is now widely adopted (R. H. Whittaker, 1966, Science *163*:150). Such a system is shown in the *phylogenetic tree* (Fig. A9.1) which gives a summary of organism classification.

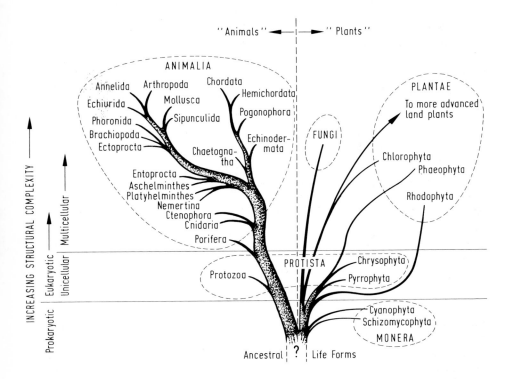

Fig A9.1. Evolutionary tree. Hypothetical evolutionary tree showing probable relationships of the major groups of organisms. Kingdoms are outlined with *dashed lines*. (J. L. Sumich, 1976, An introduction to the biology of marine life. W. C. Brown Co., Dubuque, Iowa)

Kingdom Monera: Prokaryotic cells: lacking nuclear membranes

Bacteria: Responsible for remineralization of organic matter and regeneration of nutrients in the water column and on the sea floor. Also for nitrogen fixation, sulfate and nitrate reduction, sulfide precipitation (pyrite), ferromanganese precipitation (manganese nodules). Photosynthetic bacteria in the sunlit zone add to primary production. Chemosynthetic bacteria form the base of the food chain at hydrothermal vents.

Blue-green algae (Cyanophyta): Photosynthesis in plankton and benthos. Stromatolite formation. Nitrogen fixation. Endolithic forms in calcareous shells. Chemosynthetic forms. Algalmats in lagoons.

Kingdom Protista: Unicellular eukaryotic cells, possessing nuclear membranes. There are two major groups; unicellular algae (e. g., dinoflagellates, coccolithophores, diatoms, silicoflagellates) and protozoans (e.g., foraminifera, radiolarians, and tintinnids).

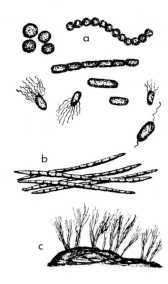

269

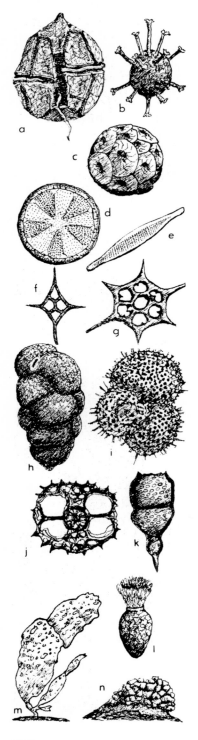

Dinoflagellates: Photosynthetic plankton, commonly the dominant form in seawater. Certain types are responsible for "red tide" which can poison fish and render mussels inedible. Common as *symbionts* ("zooxanthellae") in foraminifera, radiolarians, corals, sponges, and some mollusks. Their cysts are found as fossils in organic-rich sediments. Phylum Pyrrophyta.

Coccolithophores: photosynthetic plankton and benthos (the latter rarely as fossils). Calcitic platelets of certain pelagic species are preserved in deep-sea sediments, hence, coccoliths are probably the most abundant type of fossils on Earth (since the Mesozoic). Rockforming (e. g., Cretaceous chalks). Phylum Chrysophyta.

Diatoms: Photosynthetic plankton and benthos, extremely common in all watery environments. Siliceous skeleton (pill-box principle). Colonial forms abundant in ocean. Characteristic component of sediments in upwelling regions. Rockforming in Tertiary deposits (diatomite). Phylum Chrysophyta.

Silicoflagellates: Somewhat similar to diatoms but with an inner skeleton of silica. Photosynthetic plankton of the ocean. Abundant in fertile regions and their sediments.

Foraminifera: Heterotrophic plankton and benthos, marine only. Calcareous shells. Also agglutinated tests, on the sea floor. Rockforming (e. g., Eocene nummulitic limestones: building material of Egyptian pyramids). Foraminiferal ooze ("Globigerina ooze") is the single most widespread deposit on the face of Earth at present (covers ca. ½ of deep-sea floor). Many plankton and reef species have dinoflagellate symbionts.

Radiolarians: Somewhat similar to foraminifera, but with internal skeletons made of silica. Marine plankton only, all depths. Common in pelagic sediments below fertile regions. Rockforming (radiolarites of late Jurassic). Shallow planktonic forms can have dinoflagellate symbionts. Main groups: nasselarians (opal), spumellarians (opal), phaeodarians (organic-rich opal skeleton). Also there is a distinct group, the acantharians, which have skeletons made of $SrSO_4$. Acantharians are common in tropical surface waters, but do not make fossils.

Tintinnids: Ciliate protozoans with an organic test ("lorica"). Abundant in marine plankton. Loricae are fossilized in organic-rich sediments (upwelling regions).

Kingdom Plantae: Multicellular, with walled eukaryotic cells, photosynthetic. There are two major groups: the multicellular algae (e. g., Rhodophyta, Phaeophyta, Chlorophyta) and the higher plants (Metaphyta).

Rhodophyta (Red Algae): Common in the littoral zone, especially in warm regions, and on hard substrates down to very low light levels. Corallinaceae (e. g.,

270

Lithothamnion) have calcified cell walls, they are important sediment producers in places.

Phaeophyta (Brown Algae): Best known as kelp forests of temperate and cold water coasts, the most highly productive marine biotope. Typical representatives: *Laminaria, Macrocystis, Fucus. Sargassum* is abundant in the Sargasso Sea, where it floats on the surface. Delivery of organic sediments to coastal regions (possibly also below Sargasso Sea).

Clorophyta (Green Algae): Common at the sea shore (e. g., sea lettuce, *Ulva*). In warm waters abundant lime-secreting forms, e. g., *Halimeda* (Codiaceae) and representatives of the Dasycladaceae which were rock-forming in times past (e. g., alpine Triassic).

Metaphyta: The higher plants (mosses, vascular plants) are primarily terrestrial. Exceptions are: the sea grasses (e. g., *Zostera*) in the littoral zone, which are flowering plants. Certain halophytes (*Salicornia* in salt marshes; mangroves in tropical coastal swamps) are important in the intertidal zone.

Kingdom Fungi: Unicellular and multicellular eukaryotic organisms lacking photosynthetic pigments and depending on nutrition by absorption. There are few marine species.

Kingdom Animalia: Multicellular eukaryotic organisms without photosynthetic pigments and depending on nutrition by ingestion. Higher forms have sensory-neuro-motor systems.

Porifera (Sponges): Most species are marine. Common on firm sea floor in fertile area, in all latitudes. Boring sponges *(Cliona)* produce sediment particles, weaken mollusk shells, and erode calcareous substrates. Calcareous and siliceous form can be rock-forming. Siliceous sponge spicules accumulate in great masses on the Antarctic shelf, in places.

Cnidaria (Coelenterata) Mostly marine. Planktonic forms (hydrozoans, siphonophores, scyphozoans) and benthic forms (hydrozoans, anthozoans) abundant. Stone corals contribute greatly to reef building in tropical regions (*Acropora, Porites, Pocillopora,* etc.). Deep water and cold water stone corals also exist (e. g., *Lophelia*) but they do not build reefs.

Worms This common-name category for elongate soft-bodied animals (Vermes) no longer has scientific standing. There are several phyla of worms, such as: *Platyhelminthes,* flatworms (some free-living, but mostly parasites), *Nemertea,* ribbon worms (common on the sea floor), *Aschelminthes,* round worms (containing the important class of nematodes, which is very abundant in the marine benthos. Parasites common), *Phoronida,* phoronid worms (few species, but common), *Sipunculoidea,* peanut worms (few species),

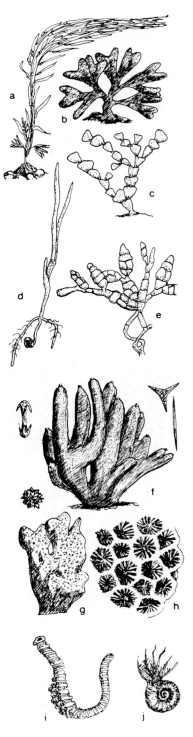

271

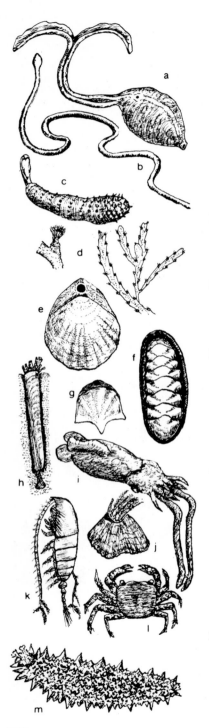

Echiuridea, spoon worms (few species, deposit feeders), *Annelida,* segmented worms (common in plankton and benthos; well-known representative: *Arenicola*), *Pogonophora,* beard worms (lack a digestive tract; tube dwellers, mainly soft bottom, deep sea, few species), *Hemichordata* (pterobranch and acorn worms; pterobranchia are common on the sea floor over a large depth range, tube-building species colony-forming in places; acorn worms are common in soft sediment, mainly on shelf).

Worms are important in reworking and processing surface sediment, and in subsurface irrigation of the sea floor (enhancing chemical exchange between sediment and water). Filter-feeding worms convert suspended matter to sediment. Tube-building worms can provide solid substrates for epibenthos to settle on.

Bryozoans (Moss Animals): Benthic colonial organisms superficially resembling small colonial hydroids. Many species secrete a a protective and supporting structure made of carbonate. These structures may be encrusting rocks or shells, or may stand as small bushes or trees.

Brachiopods (Lamp Shells): Common as fossils, but not so important in the present ocean (30,000 extinct species, 300 living ones). Benthic organisms, superficially resembling bivalves. Shells are made of chitinophosphatic or of calcareous material.

Molluscs: Highly diverse group (more than 60,000 species) and of major importance in sediment formation ($CaCO_3$). Includes gastropods (snails), pelecypods (mussels, clams, oysters, etc.), cephalopods (squids, octopuses, etc.) and the less well-known classes amphineurans (chitons), scaphopods (tusk shells) and monoplacophorans, a rare group thought to be extinct since the Paleozoic before representatives were dicovered in the deep sea in the 1950's.

Arthropods: Most diverse of the animal phyla: arthropods account for more than 75% of all animal species. The great diversity of the insects is responsible; there are only a few marine insects, which belong to the genus *Halobates,* the water strider. The arthropods include the crustaceans, merostomatans, and pycnogonids (sea spiders). The crustaceans (shrimps, crabs, lobsters, barnacles, isopods, amphipods, and others) are ubiquitous in and on the sea floor, reworking sediment and also contributing to sediment formation, by pelletization, and by precipitation of phosphatic and calcareous hard parts. The best-known representative of the merostomatans is the horse-shoe crab.

Echinoderms: Ubiquitous on the sea floor: echinoids (sea urchins), asteroids (sea stars), crinoids (sea lilies), holothurians (sea cucumbers), and ophiuroids (brittle stars). Holothurians ingest and process considerable amounts of surface sediment in certain regions of the continental slope and beyond. Virtually all echinoderms precipitate calcium carbonate, which contributes substantially to shallow water carbonates in places.

Chordates: Highly diverse phylum including the familiar backboned animals (fishes, amphibians, reptiles, birds, mammals) and also some rather primitive forms: the (benthic) sea squirts and the (planktonic) salps. Salps are important in pelletizing extremely fine-grained suspended matter, which removes it from surface waters and expedites its way to the sea floor.

Organisms illustrated:

Page 269
a spherical, rod-shaped, and flagellated bacteria
b pelagic blue-green alga (*Trichodesmium,* which is red-colored and gives the Red Sea its name)
c colony of a benthic blue-green alga *(Phormidium)*

Page 270
a, b dinoflagellate *(Gonyaulax)* and its resting cyst ("hystrichosphere")
c coccosphere *(Cyclococcolithus)* covered with coccoliths
d, e centric diatom *(Actinoptychus)* and pennate diatom *(Nitzschia)*
f, g common silicoflagellates *(Distephanus, Dictyocha)*
h, i benthic foram *(Verneuilina)* and planktonic foram *(Globigerinoides)*
j, k spumellarian and nasselarian radiolarians *(Octopyle stenozona, Theocorythium trachelium)*
l tintinnid *(Stenosemella)*
m, n red algae *(Rhodymenia, Lithothamnium)*

Page 271
a, b brown algae, giant kelp *(Macrocystis),* rockweed *(Fucus)*
c calcareous green alga *(Halimeda)*
d sea grass *(Zostera)*
e marine fungus *(Alternaria)*
f calcareous sponge *(Sycon);* also shown are various types of sponge spicules
g benthic calcareous hydrozoan (Millepora)
h stone coral *(Favites)*
i, j annelid worms (the lugworm *Arenicola, the serpulid Spirorbis*)

Page 272
a, b, c worms: echiurid *(Boniellia),* nemertine *(Lineus),* echiurid *(Echiurus)*
d bryozoan colony *(Crisia)*
e brachiopod (terebratulid)
f, g, h, i shell-forming molluscs: chiton or sea cradle *(Nuttallina),* pelagic snail *(Cavolinia),* razor clam *(Ensis),* shell-forming squid *(Spirula)*
j, k, l arthropods: cirriped (barnacle, *Balanus*), copepod *(Calanus),* shore crab *(Crassipes)*
m holothurian or sea cucumber *(Aspidochirota)*

page 273
a sea squirt *(Halocynthia)*
b sting ray *(Taemiura)*

Sources for A9: E. Y. Dawson, 1966. Marine botany, Holt Rinehart Winston, New York; B. A. Haq, H. Boersma, 1978. Introduction to marine micropaleontology, Elsevier, New York; A. Remane, V. Storch, U. Welsch, 1976. Systematische Zoologie, Gustav Fischer, Stuttgart; G. G. Simpson, W. S. Beck, 1965, Life — an introduction to biology, 2nd ed., Harcourt Brace Jovanovich, New York; H. U. Sverdrup, M. W. Johnson, R. H. Fleming, 1942, The oceans — their physics, chemistry, and general biology. Prentice-Hall, Englewood Cliffs.

Index of Names

(for additional entries, see p. 250, List of Books and Symposia)

275

Subject Index

286